Encyclopedia of Hydrodynamics: Current Topics

Volume III

Encyclopedia of Hydrodynamics: Current Topics
Volume III

Edited by **Fay McGuire**

New York

Published by NY Research Press,
23 West, 55th Street, Suite 816,
New York, NY 10019, USA
www.nyresearchpress.com

Encyclopedia of Hydrodynamics: Current Topics
Volume III
Edited by Fay McGuire

International Standard Book Number: 978-1-63238-135-4 (Hardback)

.

Printed in the United States of America.

Contents

Preface

In my initial years as a student, I used to run to the library at every possible instance to grab a book and learn something new. Books were my primary source of knowledge and I would not have come such a long way without all that I learnt from them. Thus, when I was approached to edit this book; I became understandably nostalgic. It was an absolute honor to be considered worthy of guiding the current generation as well as those to come. I put all my knowledge and hard work into making this book most beneficial for its readers.

The phenomena associated with the flow of fluids are usually complex, and tough to quantify. Novel approaches - considering points of view still not investigated - may present useful devices in the study of Hydrodynamics and the associated transport phenomenon. The specifications of the flows and the characteristics of the fluids must be studied on a small scale. Subsequently, novel concepts and devices are devised to better explain the fluids and their characteristics. This book provides conclusions about advanced issues of calculated and observed flows. Major topics discussed are mathematical models in fluid mechanics; biological operations and bio-hydrodynamics; and exploratory evaluation of fluids and flows. This book introduces novel viewpoints about procedures and tools used in Hydrodynamics.

I wish to thank my publisher for supporting me at every step. I would also like to thank all the authors who have contributed their researches in this book. I hope this book will be a valuable contribution to the progress of the field.

Editor

Part 1

Mathematical Models in Fluid Mechanics

One Dimensional Turbulent Transfer Using Random Square Waves – Scalar/Velocity and Velocity/Velocity Interactions

H. E. Schulz[1,2], G. B. Lopes Júnior[2], A. L. A. Simões[2] and R. J. Lobosco[2]
[1]Nucleus of Thermal Engineering and Fluids
[2]Department of Hydraulics and Sanitary Engineering School
of Engineering of São Carlos, University of São Paulo
Brazil

1. Introduction

The mathematical treatment of phenomena that oscillate randomly in space and time, generating the so called "statistical governing equations", is still a difficult task for scientists and engineers. Turbulence in fluids is an example of such phenomena, which has great influence on the transport of physical proprieties by the fluids, but which statistical quantification is still strongly based on *ad hoc* models. In turbulent flows, parameters like velocity, temperature and mass concentration oscillate continuously in turbulent fluids, but their detailed behavior, considering all the possible time and space scales, has been considered difficult to be reproduced mathematically since the very beginning of the studies on turbulence. So, statistical equations were proposed and refined by several authors, aiming to describe the evolution of the "mean values" of the different parameters (see a description, for example, in Monin & Yaglom, 1979, 1981).

The governing equations of fluid motion are nonlinear. This characteristic imposes that the classical statistical description of turbulence, in which the oscillating parameters are separated into mean functions and fluctuations, produces new unknown parameters when applied on the original equations. The generation of new variables is known as the "closure problem of statistical turbulence" and, in fact, appears in any phenomenon of physical nature that oscillates randomly and whose representation is expressed by nonlinear conservation equations. The closure problem is described in many texts, like Hinze (1959), Monin & Yaglom (1979, 1981), and Pope (2000), and a general form to overcome this difficulty is matter of many studies.

As reported by Schulz et al. (2011a), considering scalar transport in turbulent fluids, an early attempt to theoretically predict RMS profiles of the concentration fluctuations using "ideal random signals" was proposed by Schulz (1985) and Schulz & Schulz (1991). The authors used random square waves to represent concentration oscillations during mass transfer across the air-water interface, and showed that the RMS profile of the concentration fluctuations may be expressed as a function of the mean concentration profile. In other words, the mean concentration profile helps to know the RMS profile. In these studies, the authors did not consider the effect of diffusion, but argued that their equation furnished an upper limit for the normalized RMS value, which is not reached when diffusion is taken into account.

The random square waves were also used by Schulz et al. (1991) to quantify the so called "intensity of segregation" in the superficial boundary layer formed during mass transport, for which the explanations of segregation scales found in Brodkey (1967) were used. The time constant of the intensity of segregation, as defined in the classical studies of Corrsin (1957, 1964), was used to correlate the mass transfer coefficient across the water surface with more usual parameters, like the Schmidt number and the energy dissipation rate. Random square waves were also applied by Janzen (2006), who used the techniques of Particle Image Velocimetry (PIV) and Laser Induced Fluorescence (LIF) to study the mass transfer at the air-water interface, and compared his measurements with the predictions of Schulz & Schulz (1991) employing *ad hoc* concentration profiles. Further, Schulz & Janzen (2009) confirmed the upper limit for the normalized RMS of the concentration fluctuations by taking into account the effect of diffusion, also evaluating the thickness of diffusive layers and the role of diffusive and turbulent transports in boundary layers. A more detailed theoretical relationship for the RMS of the concentration fluctuation showed that several different statistical profiles of turbulent mass transfer may be interrelated.

Intending to present the methodology in a more organized manner, Schulz et al. (2011a) showed a way to "model" the records of velocity and mass concentration (that is, to represent them in an *a priori* simplified form) for a problem of mass transport at gas-liquid interfaces. The fluctuations of these variables were expressed through the so called "partition, reduction, and superposition functions", which were defined to simplify the oscillating records. As a consequence, a finite number of basic parameters was used to express all the statistical quantities of the equations of the problem in question. The extension of this approximation to different Transport Phenomena equations is demonstrated in the present study, in which the mentioned statistical functions are derived for general scalar transport (called here "scalar-velocity interactions"). A first application for velocity fields is also shown (called here "velocity-velocity interactions"). A useful consequence of this methodology is that it allows to "close" the turbulence equations, because the number of equations is bounded by the number of basic parameters used. In this chapter we show 1) the *a priori* modeling (simplified representation) of the records of turbulent variables, presenting the basic definitions used in the random square wave approximation (following Schulz et al., 2011a); 2) the generation of the usual statistical quantities considering the random square wave approximation (scalar-velocity interactions); 3) the application of the methodology to a one-dimensional scalar transport problem, generating a closed set of equations easy to be solved with simple numerical resources; and 4) the extension of the study of Schulz & Johannes (2009) to velocity fields (velocity-velocity interactions).

Because the method considers primarily the oscillatory records itself (*a priori* analysis), and not phenomenological aspects related to physical peculiarities (*a posteriori* analysis, like the definition of a turbulent viscosity and the use of turbulent kinetic energy and its dissipation rate), it is applicable to any phenomenon with oscillatory characteristics.

2. Scalar-velocity interactions

2.1 Governing equations for transport of scalars: Unclosed statistical set

The turbulent transfer equations for a scalar F are usually expressed as

$$\frac{\partial \overline{F}}{\partial t} + \overline{V}_i \frac{\partial \overline{F}}{\partial x_i} = \frac{\partial}{\partial x_i}\left(D_F \frac{\partial \overline{F}}{\partial x_i} - \overline{v_i f} \right) + \overline{g} \, , i = 1, 2, 3. \tag{1}$$

where $\overline{F}$ and f are the mean scalar function and the scalar fluctuation, respectively. $\overline{V}_i$ ($i =$ 1, 2, 3) are mean velocities and v_i are velocity fluctuations, t is the time, x_i are the Cartesian coordinates, $\overline{g}$ represents the scalar sources and sinks and D_F is the diffusivity coefficient of F. For one-dimensional transfer, without mean movements and generation/consumption of F, equation (1) with $x_3=z$ and $v_3=\omega$ is simplified to

$$\frac{\partial \overline{F}}{\partial t} = \frac{\partial}{\partial z}\left(D_F \frac{\partial \overline{F}}{\partial z} - \overline{\omega f} \right) \tag{2}$$

As can be seen, a second variable, given by the mean product $\overline{\omega f}$, is added to the equation of $\overline{F}$, so that a second equation involving $\overline{\omega f}$ and $\overline{F}$ is needed to obtain solutions for both variables. Additional statistical equations may be generated averaging the product between equation (1) and the instantaneous fluctuations elevated to some power (f^θ). As any new equation adds new unknown statistical products to the problem, the resulting system is never closed, so that no complete solution is obtained following strictly statistical procedures (closure problem). Studies on turbulence consider a low number of statistical equations (involving only the first statistical moments), together with additional equations based on *ad hoc* models that close the systems. This procedure seems to be the most natural choice, because having already obtained equation (2), it remains to model the new parcel $\overline{\omega f}$ *a posteriori* (that is, introducing hypotheses and definitions to solve it). An example is the combined use of the Boussinesq hypothesis (in which the turbulent viscosity/diffusivity is defined) with the Komogoroff reasoning about the relevance of the turbulent kinetic energy and its dissipation rate. The $\kappa - \varepsilon$ model for statistical turbulence is then obtained, for which two new statistical equations are generated, one of them for k and the other for ε. Of course, new unknown parameters appear, but also additional *ad hoc* considerations are made, relating them to already defined variables.

In the present chapter, as done by Schulz et al. (2011a), we do not limit the number of statistical equations based on *a posteriori* definitions for $\overline{\omega f}$. Convenient *a priori* definitions are used on the oscillatory records, obtaining transformed equations for equation (1) and additional equations. The central moments of the scalar fluctuations, $f^\theta = \left[F - \overline{F} \right]^\theta$, $\theta = 1, 2, 3,\ldots$ are considered here. For example, the one-dimensional equations for $\theta=2, 3$ and 4, are given by

$$\frac{1}{2}\frac{\partial \overline{f^2}}{\partial t} + \overline{f\omega}\frac{\partial \overline{F}}{\partial z} + \frac{1}{2}\frac{\partial \overline{f^2\omega}}{\partial z} = D_F\left(\overline{f\frac{\partial^2 f}{\partial z^2}} \right) \tag{3a}$$

$$\frac{1}{3}\frac{\partial \overline{f^3}}{\partial t} + \overline{f^2}\frac{\partial \overline{F}}{\partial t} + \overline{f^2\omega}\frac{\partial \overline{F}}{\partial z} + \frac{1}{3}\frac{\partial \overline{\omega f^3}}{\partial z} = D_F\left(\overline{f^2}\frac{\partial^2 \overline{F}}{\partial z^2} + \overline{f^2\frac{\partial^2 f}{\partial z^2}} \right) \tag{3b}$$

$$\frac{1}{4}\frac{\partial \overline{f^4}}{\partial t} + \overline{f^3}\frac{\partial \overline{F}}{\partial t} + \overline{f^3 \omega}\frac{\partial \overline{F}}{\partial z} + \frac{1}{4}\frac{\partial \overline{\omega f^4}}{\partial z} = D_F\left(\overline{f^3}\frac{\partial^2 \overline{F}}{\partial z^2} + \overline{f^3\frac{\partial^2 f}{\partial z^2}}\right) \tag{3c}$$

In this example, equation (3a) involves $\overline{F}$ and $\overline{f\omega}$ of equation (2), but adds three new unknowns. The first four equations (2) and (3 a, b, c) already involve eleven different statistical quantities: $\overline{F}$, $\overline{f^2}$, $\overline{f^3}$, $\overline{f^4}$, $\overline{f\omega}$, $\overline{f^2\omega}$, $\overline{f^3\omega}$, $\overline{f^4\omega}$, $\overline{f\frac{\partial^2 f}{\partial z^2}}$, $\overline{f^2\frac{\partial^2 f}{\partial z^2}}$, and

$\overline{f^3\frac{\partial^2 f}{\partial z^2}}$, and the "closure" is not possible. The general equation for central moments, for any θ, is given by [20]

$$\frac{1}{\theta}\frac{\partial \overline{f^\theta}}{\partial t} + \overline{f^{\theta-1}}\frac{\partial \overline{F}}{\partial t} + \overline{f^{\theta-1}\omega}\frac{\partial \overline{F}}{\partial z} + \frac{1}{\theta}\frac{\partial \overline{\omega f^\theta}}{\partial z} = D_F\left(\overline{f^{\theta-1}}\frac{\partial^2 \overline{F}}{\partial z^2} + \overline{f^{\theta-1}\frac{\partial^2 f}{\partial z^2}}\right) \tag{3d}$$

(using $\theta=1$ reproduces equation (2)).
As mentioned, the method models the records of the oscillatory variables, using random square waves. The number of equations is limited by the number of the basic parameters defined "*a priori*".

2.2 "Modeling" the records of the oscillatory variables

As mentioned in the introduction, the term "modeling" is used here as "representing in a simplified way". Following Schulz et al. (2011a), consider the function $F(z, t)$ shown in Figure 1. It represents a region of a turbulent fluid in which the scalar quantity F oscillates between two functions F_p (p=previous) and F_n (n=next) in the interval $z_1 < z < z_2$. Turbulence is assumed stationary.

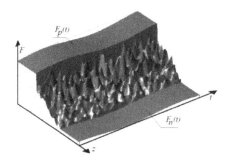

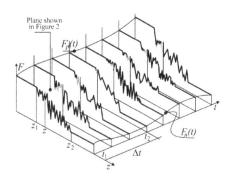

Fig. 1a. A two-dimensional random scalar field F oscillating between the boundary functions $F_p(t)$ and $F_n(t)$.

Fig. 1b. Sketch of the region shown in figure (1a). Turbulence is stationary. Adapted from Schulz et al. (2011a)

The time average of $F(z, t)$ for $z_1 < z < z_2$, indicated by $\overline{F}(z,t)$ is defined as usual

$$\overline{F}(z, t) = \frac{1}{\Delta t}\int_{t_1}^{t_2} F(z,t)dt \qquad for \qquad z_1 < z < z_2 \tag{4}$$

$\Delta t = t_2 - t_1$ is the time interval for the average operation. Equation (4) generates a mean value $\overline{F}(z)$ for $z_1 < z < z_2$ and $t_1 < t < t_2$. Any statistical quantity present in equations 3, like, for example, the central moments $\overline{f^{\theta}} = \overline{\left[F - \overline{F}\right]^{\theta}}$, is defined according to equation (4). To simplify notation, both coordinates (z, t) are dropped off in the rest of the text.

The method described in the next sections allows to obtain the relevant statistical quantities of the governing equations, like the mean function $\overline{F}$, using simplified records of F.

2.3 Bimodal square wave: Mean values using a time-partition function for the scalar field - *n*

The basic assumptions made to "model" the original oscillatory records may be followed considering Figure 2. In this sense, figure 2a is a sketch of the original record of the scalar variable F at a position $z_1 < z < z_2$, as shown in the gray vertical plane of Figure 1. The objective of this analysis is to obtain an equation for the mean function $\overline{F}(z)$ for $t_1 < t < t_2$, which is also shown in figure 2a. The values of the scalar variable during the turbulent transfer are affected by both the advective turbulent movements and diffusion. Discarding diffusion, the value of F would ideally alternate between the limits F_p and F_n (the bimodal square wave), as shown in Figure 2b (the fluid particles would transport only the two mentioned F values). This condition was assumed as a first simplification, but maintaining the correct mean, in which $\overline{F}(z)$ is unchanged. It is known that diffusion induces fluxes governed by F differences between two regions of the fluid (like the Fourier law for heat transfer and the Fick law for mass transfer). These fluxes may significantly lower the amplitude of the oscillations in small patches of fluid, and are taken into account using F_p-P and F_n+N for the two new limiting F values, as shown in Figure 2c. The parcels P and N depend on z.

In other words, the amplitude of the square oscillations is "adjusted" (modeled), in order to approximate it to the mean amplitude of the original record. As can be seen, the aim of the method is not only to evaluate $\overline{F}$ adequately, but also the lower order statistical quantities that depend on the fluctuations, which are relevant to close the statistical equations. The parcels P and N were introduced based on diffusion effects, but any cause that inhibits oscillations justifies these corrective parcels.

The first statistical parameter is represented by n, and is defined as the fraction of the time for which the system is at each of the two F values (equations 5 and 6), being thus named as "partition function". This function n depends on z and is mathematically defined as

$$n = \frac{t \ at \ (F_p - P)}{\Delta t \ of \ the \ observation} \tag{5}$$

This definition also implies that

$$1 - n = \frac{t \ at \ (F_n + N)}{\Delta t \ of \ the \ observation} \tag{6}$$

$\overline{F}$ remains the same in figures 2a, b and c. The constancy between figures 2b and c is obtained using mass conservation, implying that P and N are related through equation (7):

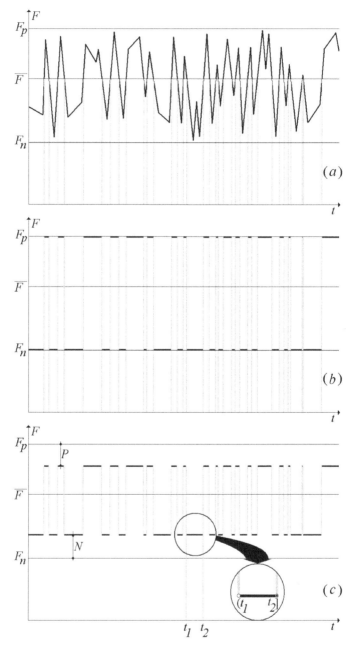

Fig. 2. *a*) Sketch of the *F* record of the gray plane of figure 1, at *z*, *b*) Simplified record alternating *F* between F_p and F_n, *c*) Simplified record with amplitude damping. Upper and lower points do not superpose at the discontinuities (the *F* segments are open at the left and closed at the right, as shown in the detail).

$$N = \frac{Pn}{(1-n)} \tag{7}$$

The mean value of F is obtained from a weighted average operation between F_p-P and F_n+N, using equations (5) through (7). It follows that

$$\overline{F} = nF_p + (1-n)F_n \tag{8}$$

Isolating n, equation (8) leads to

$$n = \frac{\overline{F} - F_n}{F_p - F_n} \tag{9}$$

Thus, the partition function n previously defined by equation (5) coincides with the normalized form of $\overline{F}$ given by equation (9). Note that n is used as weighting factor for any statistical parameter that depends on F. For example, the mean value $\overline{Q}$ of a function $Q(F)$ is calculated similarly to equation (8), furnishing

$$\overline{Q} = n\, Q\!\left(F_p - P\right) + (1-n)\, Q\!\left(F_n + N\right) \tag{9a}$$

As a consequence, equations (9) and (9a) show that any new mean function $\overline{Q}$ is related to the mean function $\overline{F}$. Or, in other words: because n is used to calculate the different mean profiles, all profiles are interrelated.

From the above discussion it may be inferred that any new variable added to the problem will have its own partition function. In the present section of scalar-velocity interactions, two partition functions are described: n for F (scalar) and m for V (velocity).

2.4 Bimodal square wave: Adjusting amplitudes using a reduction coefficient function for scalars - α_f

The sketch of figure 2c shows that the parcel P is always smaller or equal to $F_p - \overline{F}$. As already mentioned, this parcel shows that the amplitude of the fluctuations is reduced. Thus, a reduction coefficient α_f is defined here as

$$P = \alpha_f\!\left[F_p - \overline{F}\right] \qquad 0 \le \alpha_f \le 1 \tag{10}$$

where α_f is a function of z and quantifies the reduction of the amplitude due to interactions between parcels of liquid with different F values (described here as a measure of diffusion effects, but which can be a measure of any cause that inhibits fluctuations). Using the effect of diffusion to interpret the new function, values of α_f close to 1 or 0 indicate strong or weak influence of diffusion, respectively. Considering this interpretation, Schulz & Janzen (2009) reported experimental profiles for α_f in the mass concentration boundary layer during air-water interfacial mass-transfer, which showed values close to 1 in both the vicinity of the surface and in the bulk liquid, and closer to 0 in an intermediate region (giving therefore a minimum value in this region).

From equations (7), (8) and (10), N and P are now expressed as

$$\left.\begin{array}{l} N = \alpha_f\, n\left(F_p - F_n\right) \\ P = \alpha_f\left(1-n\right)\left(F_p - F_n\right) \end{array}\right\} \qquad 0 \leq \alpha_f \leq 1 \qquad (11)$$

As for the partition functions, any new variable implies in a new reduction coefficient. In the present section of scalar-velocity interactions, only the reduction coefficient for F is used (that is, α_f). In the section for velocity-velocity interactions, a reduction coefficient α_v for V (velocity) is used.

2.5 Bimodal square wave: Quantifying superposition using the superposition coefficient function - β

Let us now consider the two main variables of turbulent scalar transport, the scalar F and the velocity V, oscillating simultaneously in the interval $z_1 < z < z_2$ of Figure 1. As usual, they are represented as $F = \overline{F} + f$ and $V = \overline{V} + \omega$, where $\overline{F}$ and $\overline{V}$ are the mean values, and f and ω are the fluctuations. The correlation coefficient function $\rho(z)$ for the fluctuations f and ω is given by

$$\rho(z) = \frac{1}{\Delta t}\int_{t_1}^{t_2} \rho(z,t)\,dt = \frac{1}{\Delta t}\int_{t_1}^{t_2} \frac{\omega f}{\sqrt{\overline{\omega^2}}\sqrt{\overline{f^2}}}\,dt = \frac{\overline{\omega f}}{\sqrt{\overline{\omega^2}}\sqrt{\overline{f^2}}} \qquad (12)$$

If the fluctuations are generated by the same cause, it is expected that the records of ω and f are at least partially superposed. As done for F, it is assumed that the oscillations ω can be positive or negative and so a partition function m (a function of z) may be defined. If we consider a perfect superposition between f and ω, it would imply in $n = m$, though this is not usually the case. Aiming to consider all the cases, a superposition coefficient β is defined so that $\beta = 1.0$ reflects the direct superposition $(m = n)$, and $\beta = 0.0$ implies the inverse superposition of the positive and the negative fluctuations $(m = 1-n)$ of both fields.

The definition of β is better understood considering the scheme presented in figure 3. In this figure all positive fluctuations of the scalar variable were put together, so that the nondimensional time intervals were added, furnishing the value n. As a consequence, the nondimensional fraction of time of the juxtaposed negative fluctuations appears as $1-n$. The velocity fluctuations also appear juxtaposed, showing that $\beta = 1$ superposes f and n with the same sign (++ and --), while $\beta = 0$ superposes f and n with opposite signs (+- and -+). The positive and negative scalar fluctuations are represented by f_1 and f_2, respectively. The downwards and upwards velocity fluctuations are represented by ω_d and ω_u, respectively.

Thus, m, which defines the fraction of the time for which the system is at ω_d, is expressed as

$$m = 1 - \left(\beta + n - 2\beta n\right) \qquad (13)$$

β is a function of z. Also here any new variable implies in new superposition functions. In the present section of scalar-velocity interactions only one superposition coefficient function is used (linking scalar and velocity fluctuations).

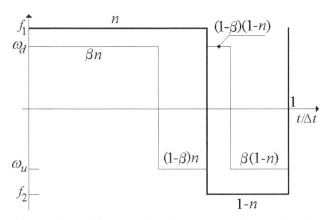

Fig. 3. Juxtaposed fluctuations of f and ω, showing a compact form of the time fractions n and $(1-n)$, and the use of the superposition function β. The horizontal axis represents the time as shown in equations (5) and (6).

2.6 The fluctuations around the mean for bimodal square waves

An advantage of using random square waves as shown in Figure 2 is that they generate only two fluctuation amplitudes for each variable, which are then used to calculate the wished statistical quantities. Of course, the functions defined in sections 2.3 through 2.5 (partition, reduction and superposition functions) are also used, and they must "adjust" the statistical quantities to adequate values. From equations (8), (10), and (11), the two instantaneous scalar fluctuations are then given by equations (14) and (15)

$$f_1 = \left(F_p - P - \overline{F}\right) = (1-n)\left(F_p - F_n\right)\left(1 - \alpha_f\right) \quad \text{(positive)} \tag{14}$$

$$f_2 = \left(F_n + N - \overline{F}\right) = -n\left(F_p - F_n\right)\left(1 - \alpha_f\right) \quad \text{(negative)} \tag{15}$$

2.7 Velocity fluctuations and the RMS velocity

In figure 1 the scalar variable is represented oscillating between two homogeneous values. But nothing was said about the velocity field that interacts with the scalar field. It may also be bounded by homogeneous velocity values, but may as well have zero mean velocities in the entire physical domain, without any evident reference velocity. This is the case, for example, of the problem of interfacial mass transfer across gas-liquid interfaces, the application shown by Schulz et al. (2011a). In such situations, it is more useful to use the rms velocity $\sqrt{\overline{\omega^2}}$ as reference, as commonly adopted in turbulence. For the one-dimensional case, with null mean motion, all equations must be derived using only the vertical velocity fluctuations ω. It is necessary, thus, to obtain equations for $\sqrt{\overline{\omega^2}}$ and for the velocity fluctuations (like equations 14 and 15 for f) considering the random square waves approximation. An auxiliary velocity scale U is firstly defined, shown in figure 4, considering "downwards" (ω_d) and "upwards" (ω_u) fluctuations, which amplitudes are functions of z.

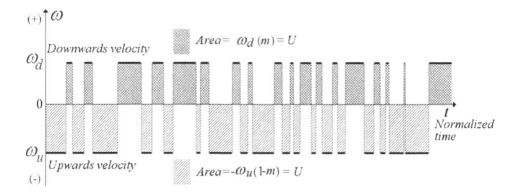

Fig. 4. The definition of the partition function m and the velocity scale U. Upwards (-) and downwards (+) velocities are shown. The dark and light gray areas are equal, so that the mean velocity is zero.

Using m for the partition function of the velocity, the scale U shown in figure 4 is defined as the integration of the upper or the lower parts of the graph in Figure 4, as

$$U = \omega_d\, m \qquad \text{and} \qquad U = -\omega_u\left(1 - m\right) \tag{16}$$

Equation (17) describes the zero mean velocity (remembering that ω_u is negative)

$$\omega_d\, m + \omega_u\left(1 - m\right) = 0 \qquad \text{or} \qquad U - U = 0 \tag{17}$$

U is a function of z. Let us now consider the RMS velocity $\sqrt{\overline{\omega^2}}$, which is calculated as

$$\overline{\omega^2} = m\,\omega_d^2 + \left(1 - m\right)\left(-\omega_u\right)^2 \qquad \text{and} \qquad \sqrt{\overline{\omega^2}} = \sqrt{m\,\omega_d^2 + \left(1 - m\right)\left(-\omega_u\right)^2} \tag{18}$$

U and $\sqrt{\overline{\omega^2}}$ may be easily related. From equations (13), (16), and (18) it follows that

$$U = \sqrt{\overline{\omega^2}}\, \sqrt{\left[1 - \left(\beta + n - 2\beta n\right)\right]\left(\beta + n - 2\beta n\right)} \tag{19}$$

Finally, the velocity fluctuations may be related to $\sqrt{\overline{\omega^2}}$, n and β using equations (16) and (19)

$$\omega_d = \sqrt{\overline{\omega^2}}\, \sqrt{\frac{\beta + n - 2\beta n}{1 - \left(\beta + n - 2\beta n\right)}} \qquad \text{and} \qquad \omega_u = -\sqrt{\overline{\omega^2}}\, \sqrt{\frac{1 - \left(\beta + n - 2\beta n\right)}{\beta + n - 2\beta n}} \tag{20}$$

$\sqrt{\overline{\omega^2}}$ is a function of z and is used as basic parameter for situations in which no evident reference velocities are present. For the example of interfacial mass transfer, $\sqrt{\overline{\omega^2}}$ is zero at the water surface ($z=0$) and constant ($\neq 0$) in the bulk liquid ($z \to \infty$).

The basic functions n, α_f, β, $\sqrt{\overline{\omega^2}}$, defined in items 2.3 through 2.7, are used in the sequence to calculate the statistical quantities of the one-dimensional equations for scalar-velocity interactions. Further, incorporating them into equations (2) and (3), a closed set of equations for these functions is generated. In other words, the one dimensional turbulent transport problem reduces to the calculation of these functions, defined *a priori* to their inclusion in the equations. Some of their general characteristics are described in table 1.

The RMS velocity may be normalized to be also bounded by the (absolute) values of 0.0 and 1.0. Because the position of the maximum value depends on the situation under study, needing more detailed explanations, the table is presented with the RMS velocity in dimensional form and having an undetermined maximum value.

Function	n	α_f	β	$\sqrt{\overline{\omega^2}}$
Dimension	Nondimensional	Nondimensional	Nondimensional	Velocity
Physical ground	Partition	Reduction	Superposition	Ref. velocity
Maximum value	1	1	1	Undetermined
Minimum value	0	0	0	0

Table 1. Characteristics of the functions defined for one dimensional scalar transport.

A further conclusion is that, because four functions need to be calculated, it implies that only four equations must be transformed to the random square waves representation in this one-dimensional situation. As a consequence, only lower order statistical quantities present in these equations need to be transformed, which is a positive consequence of this approximation, because the simplifications (and associated deviations) will not be propagated to the much higher order terms (they will not be present in the set of equations).

2.8 The central moments of scalar quantities using random square waves

It was shown that equations (3) involve central moments like $\overline{f^2}$, $\overline{f^3}$, $\overline{f^4}$, which, as mentioned, must be converted to the square waves representation. The general form of the central moments is defined as

$$\overline{f^\theta} = \overline{\left[F - \overline{F} \right]^\theta} \qquad \theta = 1, 2, 3, \ldots \qquad (21)$$

For any statistical phenomenon, the first order central moment ($\theta=1$) is always zero. Using equations (14) and (15), Schulz & Janzen (2009) showed that the second order central moment ($\overline{f^2}$ for $\theta=2$) is given by

$$\overline{f^2} = f_1^2 n + f_2^2 (1-n) = n(1-n)\left(1-\alpha_f\right)^2 \left(F_p - F_n\right)^2 \qquad (22)$$

or, normalizing the RMS value (f'_2)

$$f'_2 = \frac{\sqrt{\overline{f^2}}}{(F_p - F_n)} = \sqrt{n(1-n)}(1-\alpha_f) \qquad\qquad \alpha_f = 1 - \frac{\sqrt{\overline{f^2}}}{(F_p - F_n)\sqrt{n(1-n)}} \qquad (23)$$

This form is useful to obtain the reduction function α_f from experimental data, using the normalized mean profile and the RMS profile, as shown by Schulz & Janzen (2009). Equation (23) shows that diffusion, or other causes that inhibit the fluctuations and imply in $\alpha_f \neq 0$, imposes a peak of f'_2 lower than 0.5.

The general central moments (θ=1, 2, 3…) for the scalar fluctuation f are given by

$$\overline{f^\theta} = f_1^\theta n + f_2^\theta(1-n) = n(1-n)\left[(1-n)^{\theta-1} + (-1)^\theta (n)^{\theta-1}\right](F_p - F_n)^\theta (1-\alpha_f)^\theta \qquad (24)$$

or, normalizing the θth root (f_θ)

$$f'_\theta = \frac{\sqrt[\theta]{\overline{f^\theta}}}{(F_p - F_n)} = \sqrt[\theta]{n(1-n)\left[(1-n)^{\theta-1} + (-1)^\theta (n)^{\theta-1}\right]}(1-\alpha_f) \qquad (25)$$

The functional form of the statistical quantities shown here must be obtained solving the transformed turbulent transport equations (that is, the equations involving these quantities). Equations (21) through (25) show that, given n and α_f, it is possible to calculate all the central moments ($\overline{f^\theta}$ statistical profiles) needed in the one-dimensional equations for scalar transfer.

2.9 The covariances and correlation coefficient functions using random square waves
2.9.1 The turbulent flux of the scalar $\dot{F}$
The turbulent scalar flux, denoted by $\dot{F}$, is defined as the mean product between scalar fluctuations (f) and velocity fluctuations (ω)

$$\dot{F} = \overline{\omega f} \qquad (26)$$

Thus $\overline{\omega f}$ in equation (2) is the turbulent flux of F along z. The statistical correlation between ω and f is given by the correlation coefficient function, r, defined as

$$r = \frac{\overline{\omega f}}{\sqrt{\overline{\omega^2}}\sqrt{\overline{f^2}}} \qquad (27)$$

r is a function of z, and $0 \leq |r| \leq 1$. As it is clear from equations (26) and (27), r is also the normalized turbulent flux of F and reaches a peak amplitude less than or equal to 1.0, a range convenient for the present method, coinciding with the defined functions n, α_f, β, also bounded by 0.0 and 1.0 (as shown in table 1). The present method allows to express r as dependent on n, the normalized mean profile of F.

2.9.2 The correlation coefficient functions $\overline{f^\theta \omega}$

Equations (3) involve turbulent fluxes like $\overline{f\omega}$, $\overline{f^2\omega}$, $\overline{f^3\omega}$, $\overline{f^4\omega}$, which are unknown variables that must be expressed as functions of n, α_f, β and $\sqrt{\overline{\omega^2}}$. For products between any power of f and ω, the superposition coefficient β must be used to account for an "imperfect" superposition between the scalar and the velocity fluctuations. Therefore the flux $\overline{\omega f}$ is calculated as shown in equation (28), with β being equally applied for the positive and negative fluctuations, as shown in figure 3

$$\overline{\omega f} = \omega_d \left[f_1 n\beta + f_2(1-n)(1-\beta) \right] + \omega_u \left[f_1 n(1-\beta) + f_2(1-n)\beta \right] \tag{28}$$

Equations (13) through (20) and (28) lead to

$$\overline{\omega f} = \sqrt{\overline{\omega^2}}\left(F_p - F_n \right)\left(1-\alpha_f\right)(1-n)n(2\beta-1)\left\{ \sqrt{\frac{\beta+n-2\beta n}{1-(\beta+n-2\beta n)}} + \sqrt{\frac{1-(\beta+n-2\beta n)}{\beta+n-2\beta n}} \right\} \tag{29}$$

Rearranging, the turbulent scalar flux is expressed as

$$\overline{\omega f} = \frac{n(1-n)\left(1-\alpha_f\right)\sqrt{\overline{\omega^2}}\left(F_p - F_n \right)}{\sqrt{n(1-n)+\dfrac{\beta(1-\beta)}{(2\beta-1)^2}}} \tag{30}$$

Equations (23), (27) and (30) lead to the correlation coefficient function

$$r\Big|_{\omega,f} = \frac{\overline{\omega f}}{\sqrt{\overline{\omega^2}}\sqrt{\overline{f^2}}} = \sqrt{\frac{n(1-n)}{n(1-n)+\dfrac{\beta(1-\beta)}{(2\beta-1)^2}}} \qquad \text{with} \qquad 0 \le \left| r\Big|_{\omega,f} \right| \le 1 \tag{31}$$

Schulz el al. (2010) used this equation together with data measured by Janzen (2006). The "ideal" turbulent mass flux at gas-liquid interfaces was presented (perfect superposition of f and ω, obtained for $\beta = 1.0$). Is this case, $r\Big|_{\omega,f} = 1$, and $\overline{\omega f} = \sqrt{\overline{\omega^2}}\sqrt{\overline{f^2}}$. The measured peak of $\sqrt{\overline{\omega^2}}$, represented by W, was used to normalize $\overline{\omega f}$, as shown in Figure 5.

Considering r as defined by equation (27), it is now a function of n and β only. Generalizing for f^θ, we have

$$\overline{\omega f^\theta} = \omega_d \left[f_1^\theta n\beta + f_2^\theta (1-n)(1-\beta) \right] + \omega_u \left[f_1^\theta n(1-\beta) + f_2^\theta (1-n)\beta \right] \tag{32}$$

The correlation coefficient function is now given by

$$r\Big|_{\omega,f^\theta} = \frac{\overline{\omega f^\theta}}{\sqrt{\overline{f^{2\theta}}}\sqrt{\overline{\omega^2}}} = \sqrt{\frac{n(1-n)}{n(1-n)+\dfrac{\beta(1-\beta)}{(2\beta-1)^2}}} \left\{ \frac{\left[(1-n)^\theta - (-n)^\theta \right]}{\sqrt{\left[(1-n)^{2\theta-1} + (-1)^{2\theta}(n)^{2\theta-1} \right]}} \right\} \tag{33}$$

Fig. 5. Normalized "ideal" turbulent fluxes for $\beta=1$ using measured data. W is the measured peak of $\sqrt{\overline{\omega^2}}$. z is the vertical distance from the interface. Adapted from Schulz et al. (2011a).

Equation (32) is used to calculate covariances like $\overline{f^2\omega}$, $\overline{f^3\omega}$, $\overline{f^4\omega}$, present in equations (3). For example, for $\theta=2$, 3 and 4 the normalized fluxes are given, respectively, by:

$$r\big|_{\omega,f^2} = \frac{\overline{\omega f^2}}{\sqrt{\overline{f^4}}\sqrt{\overline{\omega^2}}} = \sqrt{\frac{n(1-n)}{n(1-n)+\dfrac{\beta(1-\beta)}{(2\beta-1)^2}}}\left\{\frac{(1-2n)}{\sqrt{\left[(1-n)^3+(n)^3\right]}}\right\} \tag{34a}$$

$$r\big|_{\omega,f^3} = \frac{\overline{\omega f^3}}{\sqrt{\overline{f^6}}\sqrt{\overline{\omega^2}}} = \sqrt{\frac{n(1-n)}{n(1-n)+\dfrac{\beta(1-\beta)}{(2\beta-1)^2}}}\left\{\frac{\left[(1-n)^3+n^3\right]}{\sqrt{\left[(1-n)^5+(n)^5\right]}}\right\} \tag{34b}$$

$$r\big|_{\omega,f4} = \frac{\overline{\omega f^4}}{\sqrt{\overline{f^8}}\sqrt{\overline{\omega^2}}} = \sqrt{\frac{n(1-n)}{n(1-n)+\dfrac{\beta(1-\beta)}{(2\beta-1)^2}}}\left\{\frac{\left[(1-n)^4-n^4\right]}{\sqrt{\left[(1-n)^7+(n)^7\right]}}\right\} \tag{34c}$$

As an ideal case, for $\beta=1$ (perfect superposition) equation 33 furnishes

$$r\big|_{\omega,f^\theta} = \frac{\overline{\omega f^\theta}}{\sqrt{\overline{f^{2\theta}}}\sqrt{\overline{\omega^2}}} = \left\{\frac{\left[(1-n)^\theta-(-n)^\theta\right]}{\sqrt{\left[(1-n)^{2\theta-1}+(-1)^{2\theta}(n)^{2\theta-1}\right]}}\right\} \tag{35}$$

and the normalized covariances $\overline{f^2\omega}$, $\overline{f^3\omega}$, $\overline{f^4\omega}$, for $\theta=2$, 3 and 4, are then given, respectively, by:

$$r\big|_{\omega,f^2} = \left\{\frac{(1-2n)}{\sqrt{\left[(1-n)^3+(n)^3\right]}}\right\} \tag{36a}$$

$$r\big|_{\omega, f^3} = \left\{ \frac{\left[(1-n)^3 + n^3\right]}{\sqrt{\left[(1-n)^5 + (n)^5\right]}} \right\} \tag{36b}$$

$$r\big|_{\omega, f4} = \left\{ \frac{\left[(1-n)^4 - n^4\right]}{\sqrt{\left[(1-n)^7 + (n)^7\right]}} \right\} \tag{36c}$$

Equations (34a) and (36a) can be used to analyze the general behavior of the flux $\overline{f^2 \omega}$. These equations involve the factor $(1 - 2n)$, which shows that this flux changes its direction at n=0.5. For 0<n<0.5 the flux $\overline{f^2 \omega}$ is positive, while for 0.5<n<1.0, it is negative. In the mentioned example of gas-liquid mass transfer, the positive sign indicates a flux entering into the bulk liquid, while the negative sign indicates a flux leaving the bulk liquid. This behavior of $\overline{f^2 \omega}$ was described by Magnaudet & Calmet (2006) based on results obtained from numerical simulations. A similar change of direction is observed for the flux $\overline{f^4 \omega}$, easily analyzed through the polynomial $(1-n)^4 - n^4$.

The equations of items 2.9.1 and 2.9.2 confirm that the normalized turbulent fluxes are expressed as functions of n and β only, while the covariances may be expressed as functions of n, β, α_f and $\sqrt{\overline{\omega^2}}$.

2.10 Transforming the derivatives of the statistical equations
2.10.1 Simple derivatives
The governing differential equations (2) and (3) involve the derivatives of several mean quantities. The different physical situations may involve different physical principles and boundary conditions, so that "particular" solutions may be found. For the example of interfacial mass transfer reported in the cited literature (e.g. Wilhelm & Gulliver, 1991; Jähne & Monahan, 1995; Donelan, et al., 2002; Janzen et al., 2010, 2011), F_p is taken as the constant saturation concentration of gas at the gas-liquid interface, and F_n is the homogeneous bulk liquid gas concentration. In this chapter this mass transfer problem is considered as example, because it involves an interesting definition of the time derivative of F_n.

The pth-order space derivative $\dfrac{\partial^p \overline{F}}{\partial z^p}$ is obtained directly from equation (8), and is given by

$$\frac{\partial^p \overline{F}}{\partial z^p} = \left(F_p - F_n\right) \frac{\partial^p n}{\partial z^p} \tag{37}$$

The time derivative of the mean concentration, $\dfrac{\partial \overline{F}}{\partial t}$, is also obtained from equation (8) and eventual previous knowledge about the time evolution of F_p and F_n. For interfacial mass transfer the time evolution of the mass concentration in the bulk liquid follows equation (38) (Wilhelm & Gulliver, 1991; Jähne & Monahan, 1995; Donelan, et al., 2002; Janzen et al., 2010, 2011)

$$\frac{d F_n}{d t} = K_f \left(F_p - F_n \right) \tag{38}$$

This equation applies to the boundary value F_n or, in other words, it expresses the time variation of the boundary condition F_n shown in figure 1. K_f is the transfer coefficient of F (mass transfer coefficient in the example). To obtain the time derivative of $\overline{F}$, equations (8) and (38) are used, thus involving the partition function n. In this example, n depends on the agitation conditions of the liquid phase, which are maintained constant along the time (stationary turbulence). As a consequence, n is also constant in time. The time derivative of $\overline{F}$ in equation (8) is then given by

$$\frac{\partial \overline{F}}{\partial t} = \frac{\partial \left[n F_p + (1-n) F_n \right]}{\partial t} = (1-n) \frac{\partial F_n}{\partial t} \tag{39}$$

From equations (38) and (39), it follows that

$$\frac{\partial \overline{F}}{\partial t} = K_f (1-n)\left(F_p - F_n \right) \tag{40}$$

Equation (40) is valid for boundary conditions given by equation (38) (usual in interfacial mass and heat transfers). As already stressed, different physical situations may conduce to different equations.

The time derivatives of the central moments $\overline{f^\theta}$ are obtained from equation (24), furnishing:

$$\frac{\partial \overline{f^\theta}}{\partial t} = -\theta n (1-n) \left[(1-n)^{\theta-1} + (-1)^\theta (n)^{\theta-1} \right] \left(F_p - F_n \right)^{\theta-1} \left(1 - \alpha_f \right)^\theta \frac{\partial F_n}{\partial t}$$

$$\text{or} \tag{41}$$

$$\frac{\partial \overline{f^\theta}}{\partial t} = -\theta K n (1-n) \left[(1-n)^{\theta-1} + (-1)^\theta (n)^{\theta-1} \right] \left(F_p - F_n \right)^\theta \left(1 - \alpha_f \right)^\theta$$

As no velocity fluctuation is involved, only the partition function n is needed to obtain the mean values of the derivatives of $\overline{f^\theta}$, that is, no superposition coefficient is needed. The obtained equations depend only on n and α_f, the basic functions related to F.

2.10.2 Mean products between powers of the scalar fluctuations and their derivatives

Finaly, the last "kind" of statistical quantities existing in equations (3) involve mean products of fluctuations and their second order derivatives, like $\overline{f \frac{\partial^2 f}{\partial z^2}}$, $\overline{f^2 \frac{\partial^2 f}{\partial z^2}}$, and $\overline{f^3 \frac{\partial^2 f}{\partial z^2}}$. The general form of such mean products is given in the sequence. From equations (14) and (15), it follows that

$$\overline{f_1^\theta \frac{\partial^2 f_1}{\partial z^2}} = \left[(1-n)\left(F_p - F_n \right)\left(1 - \alpha_f \right) \right]^\theta \frac{\partial^2 \left[(1-n)(1-\alpha_f) \right]}{\partial z^2} \left(F_p - F_n \right) \tag{42}$$

$$f_2^\theta \frac{\partial^2 f_2}{\partial z^2} = \left[-n\left(F_p - F_n\right)\left(1-\alpha_f\right)\right]^\theta \frac{\partial^2\left[-n\left(1-\alpha_f\right)\right]}{\partial z^2}\left(F_p - F_n\right) \tag{43}$$

Using the partition function n, we obtain the mean product

$$\overline{f^\theta \frac{\partial^2 f}{\partial z^2}} = \left\{(1-n)^{\theta-1}\frac{\partial^2\left[(1-n)\left(1-\alpha_f\right)\right]}{\partial z^2} + (-n)^{\theta-1}\frac{\partial^2\left[-n\left(1-\alpha_f\right)\right]}{\partial z^2}\right\}n(1-n)\left(1-\alpha_f\right)^\theta\left(F_p - F_n\right)^{\theta+1} \tag{44}$$

Equation (44) shows that mean products between powers of f and its derivatives are expressed as functions of n and α_f only.

2.11 The heat/mass transport example
In this section, the simplified example presented by Schulz et al. (2011a) is considered in more detail. The simplified condition was obtained by using a constant α_f, in the range from 0.0 to 1.0. The obtained differential equations are nonlinear, but it was possible to reduce the set of equations to only one equation, solvable using mathematical tables like Microsoft Excel® or similar.

2.11.1 Obtaining the transformed equations for the one-dimensional transport of F
Equation (2) may be transformed to its random square waves correspondent using equations (2), (8), (30), (37), and (40), leading to

$$K_f\left(1-n\right) = D_f \frac{d^2 n}{d z^2} - \frac{d}{dz}\left\{\frac{n(1-n)\left(1-\alpha_f\right)\sqrt{\overline{\omega^2}}}{\sqrt{n(1-n)+\dfrac{\beta(1-\beta)}{(2\beta-1)^2}}}\right\} \tag{45}$$

In the same way, equation (3d) is transformed to its random square waves correspondent using equations (3d), (8), (24), (32), (37), (41), and (44), leading to

$$-K_f n\left(1-n\right)\left[\left(1-n\right)^{\theta-1} + (-1)^\theta\left(n\right)^{\theta-1}\right]\left(1-\alpha_f\right)^\theta +$$

$$+K_f n\left(1-n\right)^2\left[\left(1-n\right)^{\theta-2} + (-1)^{\theta-1}\left(n\right)^{\theta-2}\right]\left(1-\alpha_f\right)^{\theta-1} +$$

$$+\sqrt{\frac{\left[n(1-n)\right]^{3-\theta}}{n(1-n)+\dfrac{\beta(1-\beta)}{(2\beta-1)^2}}}\left[\left(1-n\right)^{\theta-1} - (-n)^{\theta-1}\right]\left[n\left(1-n\right)\right]^{(\theta-1)/2}\sqrt{\overline{\omega^2}}\left(1-\alpha_f\right)^{\theta-1}\frac{\partial n}{\partial z} +$$

$$+\frac{1}{\theta}\frac{\partial}{\partial z}\left\{\sqrt{\frac{\left[n(1-n)\right]^{2-\theta}}{n(1-n)+\dfrac{\beta(1-\beta)}{(2\beta-1)^2}}}\left[\left(1-n\right)^\theta - (-n)^\theta\right]\left[n\left(1-n\right)\left(1-\alpha_f\right)^2\right]^{\theta/2}\sqrt{\overline{\omega^2}}\right\} =$$

$$= D_f n(1-n)\left[(1-n)^{\theta-2}+(-1)^{\theta-1}(n)^{\theta-2}\right](1-\alpha_f)^{\theta-1}\frac{\partial^2 n}{\partial z^2}+$$

$$+D_f\left\{(1-n)^{\theta-2}\frac{\partial^2\left[(1-n)(1-\alpha_f)\right]}{\partial z^2}+(-n)^{\theta-2}\frac{\partial^2\left[-n(1-\alpha_f)\right]}{\partial z^2}\right\}n(1-n)(1-\alpha_f)^{\theta-1} \quad (46)$$

2.11.2 Simplified case of interfacial heat/mass transfer

Although involving few equations for the present case, the set of the coupled nonlinear equations (45) and (46) may have no simple solution. As mentioned, the original one-dimensional problem needs four equations. But as the simplified solution of interfacial transfer using a mean constant $\alpha_f=\overline{\alpha_f}$ is considered here, only three equations would be

needed. Further, recognizing in equations (45) and (46) that β and $\sqrt{\omega^2}$ appear always together in the form

$$IJ=\frac{n(1-n)(1-\alpha_f)\sqrt{\omega^2}}{\sqrt{n(1-n)+\dfrac{\beta(1-\beta)}{(2\beta-1)^2}}} \quad (47)$$

It is possible to reduce the problem to a set of only two coupled equations, for n and the function IJ. Thus, only equations (45) and (46) for $\theta=2$ are necessary to close the problem when using $\alpha_f=\overline{\alpha_f}$. Defining $A=(1-\overline{\alpha_f})$ the set of the two equations is given by

$$K_f(1-n)=D_f\frac{d^2 n}{dz^2}-\frac{d(IJ)}{dz} \quad (48a)$$

$$-K_f n(1-n)A^2+(IJ)\frac{dn}{dz}+\frac{A}{2}\frac{d}{dz}\left[(IJ)(1-2n)\right]=-2D_f n(1-n)A^2\frac{d^2 n}{dz^2} \quad (48b)$$

Equations (48) may be presented in nondimensional form, using $z^*=z/E$, with $E=z_2-z_1$, and $S=1/\kappa=D_f/K_f E^2$

$$IJ^*=\frac{n(1-n)(1-\alpha_f)\left(\sqrt{\omega^2}/KE\right)}{\sqrt{n(1-n)+\dfrac{\beta(1-\beta)}{(2\beta-1)^2}}} \quad (49)$$

$$(1-n)=S\frac{d^2 n}{dz^{*2}}-\frac{d(IJ^*)}{dz^*} \quad (50a)$$

$$-n(1-n)A^2+(IJ^*)\frac{dn}{dz^*}+\frac{A}{2}\frac{d}{dz^*}\left[(IJ^*)(1-2n)\right]=-2Sn(1-n)A^2\frac{d^2 n}{dz^{*2}} \quad (50b)$$

Equation (50a) is used to obtain $dI]/dz^*$, which leads, when substituted into equation (50b), to the following governing equation for n (see appendix 1)

$$A\left[2An(1-n)+\frac{(1-2n)}{2}\right]\frac{d^3n}{dz^{*3}}\frac{dn}{dz^*}+$$

$$+A\left\{-\left[2An(1-n)+\frac{(1-2n)}{2}\right]\frac{d^2n}{dz^{*2}}+\kappa(1-n)\left[\frac{2n(A-1)+1}{2}\right]+\frac{\{1+2A[A(1-2n)-1]\}}{A}\left(\frac{dn}{dz^*}\right)^2\right\}\frac{d^2n}{dz^{*2}}+ \quad (51)$$

$$+\kappa\left\{(A-1)(1-n)-A\left[A(1-2n)-\left(\frac{3}{2}-2n\right)\right]\right\}\left(\frac{dn}{dz^*}\right)^2=0$$

Thus, the one-dimensional problem is reduced to solve equation (51) alone. It admits non-trivial analytical solution for the extreme case $A=0$ (or $\overline{\alpha_f}=1$), in the form

$$\frac{d^2n}{dz^{*2}}=\kappa(1-n) \qquad \text{or} \qquad n=1-\frac{\sin\left(\sqrt{\kappa}z^*\right)}{\sin\left(\sqrt{\kappa}\right)} \qquad (52)$$

But this effect of diffusion for all $0<z^*<1$ is considered overestimated. Equation (51) was presented by Schulz et al. (2011a), but with different coefficients in the last parcel of the first member (the parcel involving $3/2-2n$ in equation (51) involved $1-n$ in the mentioned study). Appendix 1 shows the steps followed to obtain this equation. Numerical solutions were obtained using Runge-Kutta schemes of third, fourth and fifth orders. Schulz et al. (2011a) presented a first evaluation of the n profile using a fourth order Runge-Kutta method and comparing the predictions with the measured data of Janzen (2006). An improved solution was proposed by Schulz et al. (2011b) using a third order Runge-Kutta method, in which a good superposition between predictions and measurements was obtained. In the present chapter, results of the third, fourth and fifth orders approximations are shown. The system of equations derived from (51) and solved with Runge-Kutta methods is given by:

$$\begin{cases}\frac{dn}{dz^*}=j, \quad \frac{dj}{dz^*}=w, \quad \frac{dw}{dy}=\frac{f_1+f_2}{f_3} \qquad where \\[2mm] f_1=-A\left\{-\left[2An(1-n)+\frac{(1-2n)}{2}\right]w+\kappa(1-n)\left[\frac{2n(A-1)+1}{2}\right]+\frac{\{1+2A[A(1-2n)-1]\}}{A}j^2\right\}w, \\[2mm] f_2=-\kappa\left\{(A-1)(1-n)-A\left[A(1-2n)-\left(\frac{3}{2}-2n\right)\right]\right\}j^2, \qquad\qquad\qquad (53) \\[2mm] and \\[2mm] f_3=A\left[2An(1-n)+\frac{(1-2n)}{2}\right]j. \end{cases}$$

Equations (53) were solved as an initial value problem, that is, with the boundary conditions expressed at $z^*=0$. In this case, $n(0)=1$ and $j(0)=\sim-3$ (considering the experimental data of Janzen, 2006). The value of $w(0)$ was calculated iteratively, obeying the boundary condition $0<n(1)<0.01$. The Runge-Kutta method is explicit, but iterative procedures were used to evaluate the parameters at $z^*=0$ applying the quasi-Newton method and the Solver device of the Excel® table. Appendix 2 explains the procedures followed in the table. The curves of figure 6a were obtained for $0.001 \leq \kappa \leq 0.005$, a range based on the κ experimental values of Janzen (2006), for which $\sim 0.003 < \kappa < \sim 0.004$. The values $A=0.5$ and $n''(0)=3.056$ were used to calculate n in this figure. As can be seen, even using a constant A, the calculated curve $n(z^*)$ closely follows the form of the measured curve. Because it is known that α_f is a function of z^*, more complete solutions must consider this dependence. The curve of Schulz et al. (2011a) in figure 6a was obtained following different procedures as those described here. The curves obtained in the present study show better agreement than the former one.

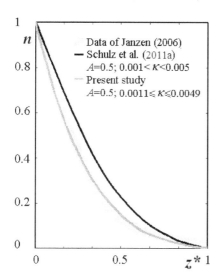

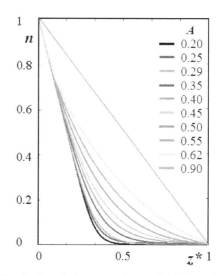

Fig. 6a. Predictions of n for $n''(0) = 3.056$. Fourth order Runge-Kutta.

Fig. 6b. Predictions of n for $\kappa = 0.0025$, and -0.0449 $\leq n''(0) \leq 3.055$. Fifth order Runge-Kutta

Fig. 6b. was obtained with following conditions for the pairs $[A, n''(0)]$: [0.2, 0.00596], [0.25, -0.0145], [0.29, -0.04495], [0.35, 1.508], [0.4, 1.8996], [0.45, 1.849], [0.5, 2.509], [0.55, 3.0547], [0.62, 2.9915], [0.90, 0.00125]. Further, $n'(0) = -3$ for A between 0.20 and 0.62, and $n'(0) = -1$ for $A=0.90$.

Figure 7a shows results for $\kappa \sim 0.4$, that is, having a value around 100 times higher than those of the experimental range of Janzen (2006), showing that the method allows to study phenomena subjected to different turbulence levels. $\kappa = (K_f E^2/D_f)$ is dependent on the turbulence level, through the parameters E and K_f, and different values of these variables allow to test the effect of different turbulence conditions on n. Figure 7b presents results similar to those of figure 6a, but using a third order Runge-Kutta method, showing that simpler schemes can be used to obtain adequate results.

As the definitions of item 3 are independent of the nature of the governing differential equations, it is expected that the present procedures are useful for different phenomena governed by statistical differential equations. In the next section, the first steps for an application in velocity-velocity interactions are presented.

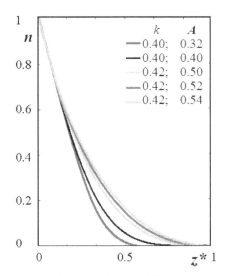

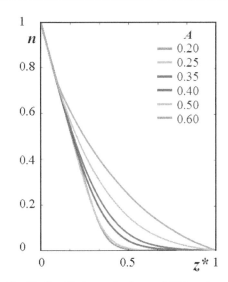

Fig. 7a. Predictions of n for $n''(0) = 3.056$, and $\kappa \sim 0.40$. Fourth order Runge-Kutta.

Fig. 7b. Predictions of n for $\kappa = 0.003$ and $2.99812 \leq n''(0) \leq 3.2111$. Third order Runge-Kutta.

3. Velocity-velocity interactions

The aim of this section is to present some first correlations for a simple velocity field. In this case, the flow between two parallel plates is considered. We follow a procedure similar to that presented by Schulz & Janzen (2009), in which the measured functional form of the reduction function is shown. As a basis for the analogy, some governing equations are first presented. The Navier-Stokes equations describe the movement of fluids and, when used to quantify turbulent movements, they are usually rewritten as the Reynolds equations:

$$\frac{\partial \overline{V_j}}{\partial t} + \overline{V_i}\frac{\partial \overline{V_j}}{\partial x_i} = \frac{\partial}{\partial x_i}\left(\upsilon\frac{\partial \overline{V_j}}{\partial x_i} - \overline{v_i v_j} \right) - \frac{1}{\rho}\frac{\partial \overline{p}}{\partial x_j} + B_i \, , \qquad i, j = 1, 2, 3. \qquad (54)$$

$\overline{p}$ is the mean pressure, υ is the kinematic viscosity of the fluid and B_i is the body force per unit mass (acceleration of the gravity). For stationary one-dimensional horizontal flows between two parallel plates, equation (1), with $x_1 = x$, $x_3 = z$, $v_1 = u$ and $v_3 = \omega$, is simplified to:

$$\frac{1}{\rho}\frac{\partial \overline{p}}{\partial x} = \frac{\partial}{\partial z}\left(\upsilon\frac{\partial \overline{U}}{\partial z} - \overline{\omega u} \right) \qquad (55)$$

This equation is similar to equation (2) for one dimensional scalar fields. As for the scalar case, the mean product $\overline{\omega u}$ appears as a new variable, in addition to the mean velocity $\overline{U}$. In this chapter, no additional governing equation is presented, because the main objective is to expose the analogy. The observed similarity between the equations suggests also to use the partition, reduction and superposition functions for this velocity field.

Both the upper and the lower parts of the flow sketched in figure 8 may be considered. We consider here the lower part, so that it is possible to define a zero velocity (U_n) at the lower surface of the flow, and a "virtual" maximum velocity (U_p) in the center of the flow. This virtual value is constant and is at least higher or equal to the largest fluctuations (see figure 8), allowing to follow the analogy with the previous scalar case.

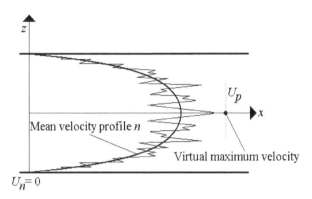

Fig. 8. The flow between two parallel planes, showing the reference velocities U_n and U_p.

The partition function n_v, for the longitudinal component of the velocity, is defined as:

$$n_v = \frac{t \ at \ (U_p - P)}{\Delta t \ of \ the \ observation} \tag{56}$$

It follows that:

$$1 - n_v = \frac{t \ at \ (U_n + N)}{\Delta t \ of \ the \ observation} \tag{57}$$

Equation (7) must be used to reduce the velocity amplitudes around the same mean velocity. It implies that the same mass is subjected to the velocity corrections P and N. As for the scalar functions, the partition function n_v is then also represented by the normalized mean velocity profile:

$$n_v = \frac{\bar{U} - U_n}{U_p - U_n} \tag{58}$$

To quantify the reduction of the amplitudes of the longitudinal velocity fluctuations, a reduction coefficient function α_u is now defined, leading, similarly to the scalar fluctuations, to:

$$\left. \begin{array}{l} N = \alpha_u \, n_v \left(U_p - U_n\right) \\ P = \alpha_u \left(1 - n_v\right)\left(U_p - U_n\right) \end{array} \right\} \qquad 0 \leq \alpha_u \leq 1 \tag{59}$$

It follows, for the x components, that:

$$u_1 = \left(1 - n_v\right)\left(U_p - U_n\right)\left(1 - \alpha_u\right) \qquad \text{(positive)} \tag{60}$$

$$u_2 = -n_v\left(U_p - U_n\right)\left(1 - \alpha_u\right) \qquad \text{(negative)} \qquad (61)$$

The second order central moment for the x component of the velocity fluctuations is given by:

$$\overline{u^2} = u_1{}^2 n_v + u_2{}^2\left(1 - n_v\right) = n_v\left(1 - n_v\right)\left(1 - \alpha_u\right)^2\left(U_p - U_n\right)^2 \qquad (62)$$

Or, normalizing the RMS value (u'_2):

$$u'_2 = \frac{\sqrt{\overline{u^2}}}{\left(U_p - U_n\right)} = \sqrt{n_v\left(1 - n_v\right)}\left(1 - \alpha_u\right) \qquad (63)$$

Equation 63 shows that the relative turbulence intensity profile is obtained from the mean velocity profile n_v and the reduction coefficient profile α_u. As done by Schulz & Janzen (2009), the profile of α_u can be obtained from experimental data, using equation (63).

$$1 - \alpha_u = \frac{\sqrt{\overline{u^2}}}{\left(U_p - U_n\right)\sqrt{n_v\left(1 - n_v\right)}} \qquad (64)$$

As can be seen, the functional form of α_u is obtainable from usual measured data, with exception of the proportionality constant given by $1/U_p$, which must be adjusted or conveniently evaluated. Figure 9 shows data adapted from Wei & Willmarth (1989), cited by Pope (2000), and the function $\sqrt{n_v\left(1 - n_v\right)}$ is calculated from the linear and log-law profiles close to the wall, also measured by Wei & Wilmarth (1989).

To obtain a first evaluation of the virtual constant velocity U_p, the following procedure was adopted. The value of the maximum normalized mean velocity is $U/u^*{\sim}24.2$ (measured), where U is the mean velocity and u^* is the shear velocity. The value of the normalized RMS u velocity, close to the peak of U, is $u'/u^*{\sim}1.14$. Considering a Gaussian distribution, 99.7% of the measured values are within the range fom U/u^*-3 u'/u^* to U/u^*+3 u'/u^*. A first value of U_p is then given by $U+3u'$, furnishing $U_p/u^*{\sim}24.2+3*1.14{\sim}27.6$. Physically it implies that patches of fluid with U_p are "transported" and reduce their velocity while approaching the wall. With this approximation, the partition function is given by:

$$n_v = \frac{u^+}{27.6} = \frac{\dfrac{1}{0.41}\ln y^+ + 5.2}{27.6} \qquad (65)$$

The value 0.41 is the von Karman constant and the value 5.2 is adjusted from the experimental data. The notation u^+ and y^+ corresponds to the nondimensional velocity and distance, respectively, used for wall flows. In this case, $u^+ = U/u^*$ and $y^+ = zu^*/v$, where v is the kinematic viscosity of the fluid. Equation (65) is the well-known logarithmic law for the velocity close to surfaces. It is generally applied for $y^+{\sim}11$. For $0<y^+{<}{\sim}11$, the linear form $u^+ = y^+$ is valid so that equation (65) is then replaced by a linear equation between n_v and y^+. From equation (63) it follows that:

$$\frac{\sqrt{\overline{u^2}}}{u^*} = \frac{U_p}{u^*}\sqrt{n_v\left(1 - n_v\right)}\left(1 - \alpha_u\right) = 27.6\sqrt{n_v\left(1 - n_v\right)}\left(1 - \alpha_u\right) \qquad (66)$$

Figure 9 shows the measured u'_2 values together with the curve given by $27.6\sqrt{n_v(1-n_v)}$. As can be seen, the curve $27.6\sqrt{n_v(1-n_v)}$ leads to a peak close to the wall. In this case, the function is normalized using the friction velocity, so that the peak is not limited by the value of 0.5 (which is the case if the function is normalized using U_p-U_n). It is interesting that the forms of $\sqrt{\overline{u'^2}}/u^*$ and $27.6\sqrt{n_v(1-n_v)}$ are similar, which coincides with the conclusions of Janzen (2006) for mass transfer, using *ad hoc* profiles for the mean mass concentration close to interfaces.

Figure 10 shows the cloud of points for $1-\alpha_u$ obtained from the data of Wei & Willmarth (1989), following the procedures of Janzen (2006) and Schulz & Janzen (2009) for mass transfer. As for the case of mass transfer, α_u presents a minimum peak in the region of the boundary layer (maximum peak for $1-\alpha_u$).

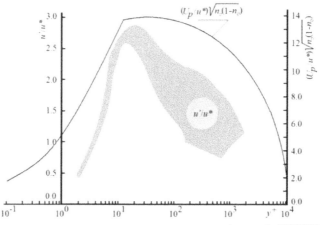

Fig. 9. Comparison between measured values of u'/u^* and $\left(U_p/u^*\right)\sqrt{n_v(1-n_v)}$. The gray cloud envelopes the data from Wei & Willmarth (1989).

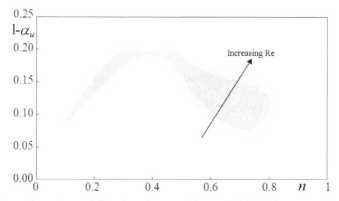

Fig. 10. $1-\alpha_u$ plotted against n, following the procedures of Schulz & Janzen (2009). The gray cloud envelopes the points calculated using the data of Wei & Willmarth (1989).

As a last observation, the conclusion of section 2.7, valid for the scalar-velocity interactions, are now also valid for the transversal component of the velocity. The mean transversal velocity is null along all the flow, leading to the use of the RMS velocity for this component.

4. Challenges

After having presented the one-dimensional results for turbulent scalar transfer using the approximation of random square waves, some brief comments are made here, about some characteristics of this approximation, and about open questions, which may be considered in future studies.

As a general comment, it may be interesting to remember that the mean functions of the statistical variables are continuous, and that, in the present approximation they are defined using discrete values of the relevant variables. As described along the paper, the defined functions (n, α, β, RMS) "adjust" these two points of view (this is perhaps more clearly explained when defining the function α). This concomitant dual form of treating the random transport did not lead to major problems in the present application. Eventual applications in 2-D, 3-D problems or in phenomena that deal with discrete variables may need more refined definitions.

In the present study, the example of mass transfer was calculated by using constant reduction coefficients (α), presenting a more detailed and improved version of the study of Schulz et al. (2011a). However, it is known that this coefficient varies along z, which may introduce difficulties to obtain a solution for n. This more complete result is still not available.

It was assumed, as usual in turbulence problems, that the lower statistical parameters (e.g. moments) are appropriate (sufficient) to describe the transport phenomena. So, the finite set of equations presented here was built using the lower order statistical parameters. However, although only a finite set of equations is needed, this set may also use higher order statistics. In fact, the number of possible sets is still "infinite", because the unlimited number of statistical parameters and related equations still exists. A challenge for future studies may be to verify if the lower order terms are really sufficient to obtain the expected predictions, and if the influence of the higher order terms alter the obtained predictions. It is still not possible to infer any behavior (for example, similar results or anomalous behavior) for solutions obtained using higher order terms, because no studies were directed to answer such questions.

In the present example, only the records of the scalar variable F and the velocity V were "modeled" through square waves. It may eventually be useful for some problems also to "model" the derivatives of the records (in time or space). The use of such "secondary records", obtained from the original signal, was still not considered in this methodology.

The problem considered in this chapter was one-dimensional. The number of basic functions for two and three dimensional problems grows substantially. How to generate and solve the best set of equations for the 2-D and 3-D situations is still unknown.

Considering the above comments, it is clear that more studies are welcomed, intending to verify the potentialities of this methodology.

5. Conclusions

It was shown that the methodology of random square waves allows to obtain a closed set of equations for one-dimensional turbulent transfer problems. The methodology adopts *a priori* models for the records of the oscillatory variables, defining convenient functions that allow to "adjust" the records and to obtain predictions of the mean profiles. This is an alternative procedure in relation to the *a posteriori* "closures" generally based on *ad hoc* models, like the use of turbulent diffusivities/viscosities, together with physical/phenomenological reasoning about relevant parameters to be considered in these diffusivities/viscosities. The basic functions are: the partition functions, the reduction coefficients and the superposition coefficients. The obtained transformed equations for the one-dimensional turbulent transport allow to obtain predictions of these functions.

In addition, the RMS of the velocity was also used as a basic function. The equations are nonlinear. An improved analysis of the one-dimensional scalar transfer through air-water interfaces was presented, leading to mean curves that superpose well with measured mean concentration curves for gas transfer. In this analysis, different constant values were used for α, κ and the second derivative at the interface, allowing to obtain well behaved and realistic mean profiles. Using the constant α values, the system of equations for one-dimensional scalar turbulent transport could be reduced to only one equation for n; in this case, a third order differential equation. In the sequence, a first application of the methodology to velocity fields was made, following the same procedures already presented in the literature for mass concentration fields. The form of the reduction coefficient function for the velocity fluctuations was calculated from measured data found in the literature, and plotted as a function of n, generating a cloud of points. As for the case of mass transfer, α_u presents a minimum peak in the region of the boundary layer (maximum peak for $1-\alpha_u$).

Because this methodology considers *a priori* definitions, applied to the records of the random parameters, it may be used for different phenomena in which random behaviors are observed.

6. Acknowledgements

The first author thanks: 1) Profs. Rivadavia Wollstein and Beate Frank (Universidade Regional de Blumenau), and Prof. Nicanor Poffo, (Conjunto Educacional Pedro II, Blumenau), for relevant advises and 2) "Associação dos Amigos da FURB", for financial support.

7. Appendix I: Obtaining equation (51)

The starting point is the set of equations (45), (46), and the definition (47).
The "*" was dropped from z^* and IJ^* in order to simplify the representation of the equations. The main equation (45) (or 50a) then is written as

$$(1-n) = S\frac{d^2 n}{dz^2} - \frac{d(IJ)}{dz} \tag{AI-1}$$

Equation (46), for $\theta=2$, is presented as:

$$-K_f n(1-n)(1-\alpha_f)^2 + \sqrt{\frac{[n(1-n)]}{n(1-n)+\frac{\beta(1-\beta)}{(2\beta-1)^2}}}[n(1-n)]^{1/2}\sqrt{\omega^2}\left(1-\alpha_f\right)\frac{\partial n}{\partial z} +$$

$$+\frac{1}{2}\frac{\partial}{\partial z}\left\{\sqrt{\frac{1}{n(1-n)+\frac{\beta(1-\beta)}{(2\beta-1)^2}}}[(1-2n)]\left[n(1-n)(1-\alpha_f)^2\right]\sqrt{\omega^2}\right\} = \tag{AI-2}$$

$$= D_f\left\{\frac{\partial^2\left[(1-n)(1-\alpha_c)\right]}{\partial z^2} + \frac{\partial^2\left[-n(1-\alpha_c)\right]}{\partial z^2}\right\}n(1-n)(1-\alpha_f)$$

Using the definitions $IJ = \dfrac{n(1-n)(1-\alpha_f)\dfrac{\sqrt{\omega^2}}{Ke}}{\sqrt{n(1-n)+\dfrac{\beta(1-\beta)}{(2\beta-1)^2}}}$ and $S = \dfrac{D}{Ke^2}$:

$$-n(1-n)(1-\alpha_c)^2 + IJ\frac{\partial n}{\partial z} + \frac{1}{2}\frac{\partial}{\partial z}\{(1-2n)(1-\alpha_c)IJ\} =$$
$$= S\left\{\frac{\partial^2\left[(1-n)(1-\alpha_c)\right]}{\partial z^2} + \frac{\partial^2\left[-n(1-\alpha_c)\right]}{\partial z^2}\right\}n(1-n)(1-\alpha_c)$$

(AI-3)

For α_f constant and defining $A=(1-\alpha_f)$:

$$-n(1-n)A^2 + IJ\frac{dn}{dz} - IJ\,A\frac{dn}{dz} + \frac{(1-2n)}{2}A\frac{dIJ}{dz} = -2S\left\{\frac{d^2n}{dz^2}\right\}n(1-n)A^2$$

(AI-4)

Using equations (AI1) and (AI4)

$$-n(1-n)A^2 - \frac{(1-n)(1-2n)}{2}A + IJ(1-A)\frac{dn}{dz} =$$
$$= -2S\left\{\frac{d^2n}{dz^2}\right\}n(1-n)A^2 - S\frac{(1-2n)}{2}A\left\{\frac{d^2n}{dz^2}\right\}$$

(AI-5)

Solving equation (AI5) for IJ:

$$IJ = \frac{-2S\left\{\dfrac{d^2n}{dz^2}\right\}n(1-n)A^2 - S\dfrac{(1-2n)}{2}A\left\{\dfrac{d^2n}{dz^2}\right\} + n(1-n)A^2 + \dfrac{(1-n)(1-2n)}{2}A}{(1-A)\dfrac{dn}{dz}}$$

(AI-6)

Rearranging equation (AI6):

$$\frac{(1-A)}{A}IJ = \frac{-S\left\{2An(1-n)+\dfrac{(1-2n)}{2}\right\}\left\{\dfrac{d^2n}{dz^2}\right\} + (1-n)\left[\dfrac{2n(A-1)+1}{2}\right]}{\dfrac{dn}{dz}}$$

(AI-7)

Differentiating equation (AI7) and using equation (AI1):

$$\frac{(1-A)}{A}\left[S\frac{d^2n}{dz^2} - (1-n)\right] =$$
$$= \frac{-S\left\{2A\left[\dfrac{dn}{dz} - 2n\dfrac{dn}{dz}\right] - \dfrac{dn}{dz}\right\}\left\{\dfrac{d^2n}{dz^2}\right\} - S\left\{2An(1-n)+\dfrac{(1-2n)}{2}\right\}\left\{\dfrac{d^3n}{dz^3}\right\} +}{\dfrac{dn}{dz}}$$

$$+\dfrac{\left(-\dfrac{dn}{dz}\right)\left[\dfrac{2n(A-1)+1}{2}\right]+(1-n)\left[\dfrac{(A-1)dn}{dz}\right]}{\dfrac{dn}{dz}}-$$

$$-\dfrac{-S\left\{2An(1-n)+\dfrac{(1-2n)}{2}\right\}\left\{\dfrac{d^2n}{dz^2}\right\}+(1-n)\left[\dfrac{2n(A-1)+1}{2}\right]}{\left(\dfrac{dn}{dz}\right)^2}\left\{\dfrac{d^2n}{dz^2}\right\}$$

(AI-8)

Multiplying by $\left(\dfrac{dn}{dz}\right)^2$ and simplifying $\dfrac{dn}{dz}$:

$$\dfrac{(1-A)}{A}\left[S\dfrac{d^2n}{dz^2}-(1-n)\right]\left(\dfrac{dn}{dz}\right)^2 =$$

$$=-\left(\dfrac{dn}{dz}\right)^2 S\{2A[1-2n]-1\}\left\{\dfrac{d^2n}{dz^2}\right\}-$$

$$-\left(\dfrac{dn}{dz}\right)S\left\{2An(1-n)+\dfrac{(1-2n)}{2}\right\}\left\{\dfrac{d^3n}{dz^3}\right\}+$$

(AI-9)

$$+\left(\dfrac{dn}{dz}\right)^2\left\{-\left[\dfrac{2n(A-1)+1}{2}\right]+(1-n)(A-1)\right\}-$$

$$-\left\{-S\left\{2An(1-n)+\dfrac{(1-2n)}{2}\right\}\left\{\dfrac{d^2n}{dz^2}\right\}+(1-n)\left[\dfrac{2n(A-1)+1}{2}\right]\right\}\left\{\dfrac{d^2n}{dz^2}\right\}$$

Rearranging (after multiplying the equation by A and using $S=1/\kappa$):

$$A\left[2An(1-n)+\dfrac{(1-2n)}{2}\right]\dfrac{d^3n}{dz^3}\dfrac{dn}{dz}+$$

$$+A\left\{\begin{array}{l}-\left[2An(1-n)+\dfrac{(1-2n)}{2}\right]\dfrac{d^2n}{dz^2}+\kappa(1-n)\left[\dfrac{2n(A-1)+1}{2}\right]+\\[2mm] +\dfrac{\{1+2A[A(1-2n)-1]\}}{A}\left(\dfrac{dn}{dz}\right)^2\end{array}\right\}\dfrac{d^2n}{dz^2}+$$

(AI-10)

$$+\kappa\left\{(A-1)(1-n)-A\left[A(1-2n)-\left(\dfrac{3}{2}-2n\right)\right]\right\}\left(\dfrac{dn}{dz}\right)^2$$

$$=0$$

Equation (AI10) is the equation (51) presented in the text.

8. Appendix II: Solving equation (51) using mathematical tables

Equation (51) (or equation (AI-10)) of this chapter is a third order nonlinear ordinary differential equation, for which adequate numerical methods must be applied. Some methods were considered to solve it.

A first attempt was made using the second order Finite Differences Method and the solver device from the Microsoft Excel® table, intending to solve the problem with simple and practical tools, but the results were not satisfactory. It does not imply that the Finite Differences Method does not apply, but only that we wanted more direct ways to check the applicability of equation (51).

The second attempt was made using Runge-Kutta methods, also furnished in mathematical tables like Excel ®, maintaining the objective of solving the one-dimensional problem with simple tools. In this case, the results were adequate, superposing well the experimental data.

The Runge-Kutta methods were developed for ordinary differential equations (ODEs) or systems of ODEs. Equation (AI-10) is a nonlinear differential equation, so that it was necessary to first rewrite it as a system of ODEs, as follows

$$\frac{dn}{dz} = j \tag{AII-1}$$

$$\frac{d^2n}{dz^2} = w \tag{AII-2}$$

$$\frac{dw}{dz} = (f_1 + f_2)/f_3 \tag{AII-3}$$

in which

$$f_1 = -A \left\{ \begin{array}{l} -\left[2An(1-n) + \dfrac{(1-2n)}{2}\right]w + \kappa(1-n)\left[\dfrac{2n(A-1)+1}{2}\right] + \\ + \dfrac{\{1+2A[A(1-2n)-1]\}}{A}j^2 \end{array} \right\} w \tag{AII-4}$$

$$f_2 = -\kappa \left\{ (A-1)(1-n) - A\left[A(1-2n) - \left(\frac{3}{2} - 2n\right)\right]\right\} j^2 \tag{AII-5}$$

$$f_3 = A\left[2An(1-n) + \frac{(1-2n)}{2}\right] j \tag{AII-6}$$

Figure 6 shows that 3th, 4th and 5th orders Runge-Kutta methods were applied to obtain numerical results for the profile of n. This Appendix shows a summary of the use of the 5th order method. Of course, similar procedures were followed for the lower orders. As usual in this chapter, equations (AII-1) up to (AII-3) use the nondimensional variable z without the star "*" (that is, it corresponds to z*). Considering "y" the dependent variable in a given ODE, the of 5th order method, presented by Butcher (1964) appud Chapra and Canale (2006), is written as follows

$$y_{k+1} = y_k + \frac{\Delta x}{90}(7\psi_1 + 32\psi_3 + 12\psi_4 + 32\psi_5 + 7\psi_6) \tag{AII-7}$$

in which

$$
\begin{cases}
\psi_1 = f\left(x_k, y_k\right) \\[6pt]
\psi_2 = f\left(x_k + \dfrac{1}{4}\Delta x, y_k + \dfrac{1}{4}\psi_1\Delta x\right) \\[6pt]
\psi_3 = f\left(x_k + \dfrac{1}{4}\Delta x, y_k + \dfrac{1}{8}\psi_1\Delta x + \dfrac{1}{8}\psi_2\Delta x\right) \\[6pt]
\psi_4 = f\left(x_k + \dfrac{1}{2}\Delta x, y_k - \dfrac{1}{2}\psi_2\Delta x + \psi_3\Delta x\right) \\[6pt]
\psi_5 = f\left(x_k + \dfrac{3}{4}\Delta x, y_k + \dfrac{3}{16}\psi_1\Delta x + \dfrac{9}{16}\psi_4\Delta x\right) \\[6pt]
\psi_6 = f\left(x_k + \Delta x, y_k - \dfrac{3}{7}\psi_1\Delta x + \dfrac{2}{7}\psi_2\Delta x + \dfrac{12}{7}\psi_3\Delta x - \dfrac{12}{7}\psi_4\Delta x + \dfrac{8}{7}\psi_5\Delta x\right)
\end{cases}
\tag{AII-8}
$$

In the system of equations (AII-8), generated from equations (AII-4) through (AII-6), $x = z$ and $y = n$, following the representation used in this chapter.

The system of equations (AII-1) through (AII-6) was solved using a spreadsheet for Microsoft Excel®, available at www.stoa.usp.br/hidraulica/files/. Two initial values were fixed and one was calculated. Note that in the present study it was intended to verify if the method furnishes a viable profile, so that boundary or initial values obtained from the experimental data were assumed as adequate. The first was $n(0)=1$. The second was $n'(0)=-3$, corresponding to the experiments of Janzen (2006). The third information did not constitute an initial value, and was $n(1)=0$ or $0<n(1)<0.01$ (threshold value corresponding to the definition of the boundary layer). As the Runge-Kutta methods need initial values, this information was used to obtain $n''(0)$, the remaining initial value needed to perform the calculations. With the aid of the Newton (or quasi-Newton) method, it was possible to obtain values for $n''(0)$ that satisfied the third condition imposed at $z = 1$.

The derivative of n at $z=0$ is generally unknown in such mass transfer problems. In this case, solutions must be found considering, for example, $n(0)=1$, $0<n(1)<0.01$ and $n'(1)=0$ (three reasonable boundary conditions), for which another scheme must be developed to calculate the first and second derivatives at the origin. As mentioned, the aim of this study was to verify the applicability of the method. The details of solutions for different purposes must be considered by the researchers interested in that solution.

The construction of the spreadsheet is described in the following steps:

i.   determine the initial values: $n(0) = 1$, $n'(0) = -3$ (or other appropriate value) $n''(0) =$ initial guess;

ii.   Compute $\psi_{1,1}$ and $\psi_{1,2}$, the function values f_1, f_2 e f_3 with the initial values, and then $\psi_{1,3}$. In the variable $\psi_{i,j}$, $i = 1,2,...,6$ and $j = 1,2,3$, the first index corresponds to the six stages of the method and the second to the order of the ODE that generated the original system to be solved;

iii. With the values calculated in (ii), calculate now $n_k+(1/4)\,\psi_{1,1}\,\Delta z$, $j_k+(1/4)\,\psi_{1,1}\,\Delta z$ and $w_k+(1/4)\,\psi_{1,1}\,\Delta z$. The following steps are similar until $j = 6$;

iv.   Equation AII-7 (a system) is then used to advance in space z.

The spreadsheet available at www.stoa.usp.br/hidraulica/files/ presents some suggestions that simplify some items of the above described steps (some manual work is simplified). The estimate of n″(0), for example, is obtained following simplified procedures.

9. References

Brodkey, R.S. (1967) The phenomena of Fluid Motions, *Addison–Wesley Publishing Company*, Reading, Massachusetts.

Butcher, J.C. (1964). On Runge-Kutta methods of high order. J.Austral. Math. Soc.4, p.179-194.

Chapra, S.C.; Canale, R.P. (2006). Numerical methods for engineers. McGraw-Hill, 5th ed., 926 p.

Corrsin, S. (1957) Simple theory of an idealized turbulent mixer, AIChE J., 3(3), pp. 329-330.

Corrsin, S. (1964) The isotropic turbulent mixer: part II - arbitrary Schmidt number, AIChE J., 10(6), pp. 870-877.

Donelan, M.A., Drennan, W.M., Saltzman, E.S. & Wanninkhof, R. (2002) Gas Transfer at Water Surfaces, Geophysical Monograph Series, American Geophysical Union, Washington, U.S.A., 383 p.

Hinze, J.O. (1959), Turbulence, Mc. Graw-Hill Book Company, USA, 586 p.

Jähne, B. & Monahan, E.C. (1995) Air-Water Gas Transfer, Selected papers from the Third International Symposium on Air-Water Gas Transfer, Heidelberg, Germany, AEON Verlag & Studio, 918 p.

Janzen, J.G. (2006) Fluxo de massa na interface ar-água em tanques de grades oscilantes e detalhes de escoamentos turbulentos isotrópicos (Gas transfer near the air-water interface in an oscillating-grid tanks and properties of isotropic turbulent flows – text in Portuguese). Doctoral thesis, University of Sao Paulo, São Carlos, Brazil.

Janzen, J.G., Herlina,H., Jirka, G.H., Schulz, H.E. & Gulliver, J.S. (2010), Estimation of Mass Transfer Velocity based on Measured Turbulence Parameters, AIChE Journal, V.56, N.8, pp. 2005-2017.

Janzen J.G, Schulz H.E. & Jirka GH. (2006) Air-water gas transfer details (portuguese). Revista Brasileira de Recursos Hídricos; 11, pp. 153-161.

Janzen, J.G., Schulz, H.E. & Jirka, G.H. (2011) Turbulent Gas Flux Measurements near the Air-Water Interface in an Oscillating-Grid Tank. In Komori, S; McGillis, W. & Kurose, R. Gas Transfer at Water Surfaces 2010, Kyoto University Press, Kyoto, pp. 65-77.

Monin, A.S. & Yaglom, A.M. (1979), Statistical Fluid Mechanics: Mechanics of Turbulence, Volume 1, the MIT Press, 4th ed., 769p.

Monin, A.S. & Yaglom, A.M. (1981), Statistical Fluid Mechanics: Mechanics of Turbulence, Volume 2, the MIT Press, 2th ed., 873p.

Pope, S.B. (2000), Turbulent Flows, Cambridge University Press, 1st ed., UK, 771p.

Schulz, H.E. (1985) Investigação do mecanismo de reoxigenação da água em escoamento e sua correlação com o nível de turbulência junto à superfície - 1. (Investigation of the reoxigenation mechanism in flowing waters and its relation to the turbulence level at the surface-1 – text in Portuguese) MSc dissertation, University of São Paulo, Brazil São Carlos. 299p.

Schulz, H.E.; Bicudo, J.R., Barbosa, A.R. & Giorgetti, M.F. (1991) Turbulent Water Aeration: Analytical Approach and Experimental Data, In Wilhelms, S.C. and Gulliver, J.S., eds. Air Water Mass Transfer, ASCE, New York, pp.142-155.

Schulz, H.E. & Janzen, J.G. (2009) Concentration fields near air-water interfaces during interfacial mass transport: oxygen transport and random square wave analysis. *Braz. J. Chem. Eng.* vol.26, n.3, pp. 527-536.

Schulz, H.E., Lopes Junior, G.B. & Simões, A.L.A. (2011b) Gas-liquid mass transfer in turbulent boundary layers using random square waves, 3rd workshop on fluids and PDE, June 27 to July 1, Institute of Mathematics, Statistics and Scientific Computation, Campinas, Brazil.

Schulz H.E. & Schulz S.A.G. (1991) Modelling below-surface characteristics in water reaeration. Water pollution, modelling, measuring and prediction. Computational Mechanics Publications and Elsevier Applied Science, pp. 441–454.

Schulz, H.E., Simões, A.L.A. & Janzen, J.G. (2011a), Statistical Approximations in Gas-Liquid Mass Transfer, In Komori, S; McGillis, W. & Kurose, R. Gas Transfer at Water Surfaces 2010, Kyoto University Press, Kyoto, pp. 208-221.

Wilhelms, S.C. & Gulliver, J.S. (1991) Air-Water Mass Transfer, Selected Papers from the Second International Symposium on Gas Transfer at Water Surfaces, Minneapolis, U.S.A., ASCE, 802 p.

Planar Stokes Flows with Free Boundary

Sergey Chivilikhin[1] and Alexey Amosov[2]
*[1]National Research University of Information
Technologies, Mechanics and Optics,
[2]Corning Scientific Center, Corning Incorporated
Russia*

1. Introduction

The quasi-stationary Stokes approximation (Frenkel, 1945; Happel & Brenner, 1965) is used to describe viscous flows with small Reynolds numbers. Two-dimensional Stokes flow with free boundary attracted the attention of many researches. In particular, an analogy is drawn (Ionesku, 1965) between the equations of the theory of elasticity (Muskeleshvili, 1966) and the equations of hydrodynamics in the Stokes approximation. This idea allowed (Antanovskii, 1988) to study the relaxation of a simply connected cylinder under the effect of capillary forces. Hopper (1984) proposed to describe the dynamics of the free boundary through a family of conformal mappings. This approach was later used in (Jeong & Moffatt, 1992; Tanveer & Vasconcelos, 1994) for analysis of free-surface cusps and bubble breakup.

We have developed a method of flow calculation, which is based on the expansion of pressure in a complete system of harmonic functions. The structure of this system depends on the topology of the region. Using the pressure distribution, we calculate the velocity on the boundary and investigate the motion of the boundary. In case of capillary forces the pressure is the projection of a generalized function with the carrier on the boundary on the subspace of harmonic functions (Chivilikhin, 1992).

We show that in the 2D case there exists a non-trivial variation of pressure and velocity which keeps the Reynolds stress tensor unchanged. The correspondent variations of pressure give us the basis for pressure presentation in form of a series. Using this fact and the variation formulation of the Stokes problem we obtain a system of equations for the coefficients of this series. The variations of velocity give us the basis for the vortical part of velocity presentation in the form of a serial expansion with the same coefficients as for the pressure series.

We obtain the potential part of velocity on the boundary directly from the boundary conditions - known external stress applied to the boundary. After calculating velocity on the boundary with given shape we calculate the boundary deformation during a small time step.

Based on this theory we have developed a method for calculation of the planar Stokes flows driven by arbitrary surface forces and potential volume forces. We can apply this method for investigating boundary deformation due to capillary forces, external pressure, centrifugal forces, etc.

Taking into account the capillary forces and external pressure, the strict limitations for motion of the free boundary are obtained. In particular, the lifetime of the configurations with given number of bubbles was predicted.

2. General equations

2.1 The quasi-stationary Stokes approximation

The equations of viscous fluid motion in the quasi-stationary Stokes approximation due to arbitrary surface force f_α and the continuity equation in the region $G \subset R^2$ with boundary Γ have the form

$$\frac{\partial p_{\alpha\beta}}{\partial x_\beta} = 0 , \tag{1}$$

$$\frac{\partial v_\beta}{\partial x_\beta} = 0 , \tag{2}$$

where $p_{\alpha\beta} = -p\delta_{\alpha\beta} + \mu\left(\dfrac{\partial v_\alpha}{\partial x_\beta} + \dfrac{\partial v_\beta}{\partial x_\alpha}\right)$ is the Newtonian stress tensor; v_α are the components

of the velocity; p is the pressure; μ is the coefficient of the dynamical viscosity, which is assumed to be constant. The indices α, β take the values 1, 2. Summation over repeated indices is expected. The boundary conditions have the form

$$p_{\alpha\beta}n_\beta = f_\alpha, \quad \mathbf{x} \in \Gamma \tag{3}$$

where n_α and f_α are the components of the vector of outer normal to the boundary and the surface force. Let Γ_0 be the outer boundary of the region; $\Gamma_k (k = 1,2,...,m)$ - the inner

boundaries (boundaries of bubbles); $\Gamma = \bigcup\limits_{k=0}^{m} \Gamma_k$ - see Fig.1.

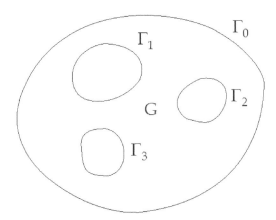

Fig. 1. Region G with multiply connected boundary Γ

The free boundary evolution is determined from the condition of equality of the normal velocity V_n of the boundary and the normal component of the velocity of the fluid at the boundary:

$$V_n = v_\beta n_\beta, \quad \mathbf{x} \in \Gamma \tag{4}$$

In case of a volume force F_α acting on G, the equation of motion takes the form

$$\frac{\partial p_{\alpha\beta}}{\partial x_\beta} = -F_\alpha \tag{5}$$

If the volume force is potential $F_\alpha = -\dfrac{\partial U}{\partial x_\alpha}$ one can renormalize the pressure $p \to p + U$ and present (3), (5) in the form

$$\frac{\partial p_{\alpha\beta}}{\partial x_\beta} = 0 \tag{6}$$

$$p_{\alpha\beta} n_\beta = f'_\alpha, \quad \mathbf{x} \in \Gamma \tag{7}$$

where $f'_\alpha = f_\alpha + U n_\alpha$ is the renormalized surface force.

2.2 The transformational invariance of the Stokes equations
Let's point out a specificity of the quasi-stationary Stokes approximation (1), (2). This system is invariant under the transformation

$$v_\alpha \to v_\alpha + V_\alpha + e_{\alpha\beta} x_\beta \omega \tag{8}$$

where V_α and ω are constants, $e_{\alpha\beta}$ is the unit antisymmetric tensor. Therefore, for this approximation the total linear momentum and the total angular momentum are indefinite. These values should be determined from the initial conditions.

2.3 The conditions of the quasi-stationary Stokes approximation applicability
The Navier-Stokes equations

$$\rho\left(\frac{\partial v_\alpha}{\partial t} + v_\beta \frac{\partial v_\alpha}{\partial x_\beta}\right) = \frac{\partial p_{\alpha\beta}}{\partial x_\beta} + F_\alpha, \tag{9}$$

where ρ is the density of liquid, lead to the quasi-stationary Stokes equations (5) if the convective and non-stationary terms in (9) can be neglected. The neglection of the convective term leads to the requirement of a small Reynolds number $\text{Re} = VL/v$, where V is the characteristic velocity, L is the spatial scale of the region G, and v is the kinematic viscosity. The non-stationary term in the equation (9) can be omitted if during the velocity field relaxation time $T = L^2/v$ the shape of the boundary changes insignificantly, namely $VT \ll L$ which again leads to the condition $\text{Re} \ll 1$. The change of the volume force F_α and the surface force f_α during the time T should also be small:

$$\frac{\delta F_\alpha}{\delta t} T \ll F_a, \quad \frac{\delta f_\alpha}{\delta t} T \ll f_a, \tag{10}$$

For the forces determined by the region shape (like capillary force or centrifugal force) the conditions (10) lead to $\mathrm{Re} \ll 1$ again.

The neglection of the non-stationary term is a singular perturbation of the motion equation in respect of the time variable. It leads to the formation of a time boundary layer of duration T, during which the initial velocity field relaxates to a quasi-steady state. The condition of a small deformation of the region during this time interval $V^0 T \ll L^0$ is ensured by the requirement of a small Reynolds number Re^0 constructed from the characteristic initial velocity V^0 and the initial region scale L^0.

Let's integrate the motion equation (5) over the region G and use the boundary condition (3). As a result we obtain the condition

$$\Phi_\alpha = \int F_\alpha dG + \int f_\alpha d\Gamma = 0. \tag{11}$$

The equations of viscous fluid motion in the quasi-stationary Stokes approximation (5) have the form of local equilibrium conditions. Correspondingly, the total force Φ_α which acts on the system should be zero. The same way, using (5) and (3) one can obtain the condition

$$M = \int e_{\alpha\beta} x_\alpha F_\beta dG + \int e_{\alpha\beta} x_\alpha f_\beta d\Gamma = 0. \tag{12}$$

where $e_{\alpha\beta}$ is the unit antisymmetric tensor. Therefore, the total moment of force M acting on the system should be zero.

2.4 The Stokes equations in the special noninertial system of reference

Conditions (11) and (12) are the classical conditions of solubility of system (2), (5) with boundary conditions (3). Let's show that these conditions are too restrictive. For example, for a small drop of high viscous liquid falling in the gravitation field the total force is not zero, but equal to the weight of the drop. Therefore, we cannot use the quasi-stationary Stokes approximation to describe the evolution of the drop's shape due to capillary forces. But in a noninertial system of reference which falls together with the drop with the same acceleration, the total force is equal to zero.

In a general case, the total force Φ_α and total moment of force M acting on the system are not equal to zero. The Newton's second law for translational motion has the form

$$\rho S \frac{d\langle v_\alpha \rangle}{dt} = \Phi_\alpha, \tag{13}$$

where S is the area of the region, $\langle v_\alpha \rangle = \frac{1}{S}\int v_\alpha dG$ is the average velocity of the system, and Φ_α is the total force. Let's choose the center-of-mass reference system K' instead of the initial laboratory system K. The velocity and coordinate transformations have the form

$$v'_\alpha = v_\alpha - \langle v_\alpha \rangle, \quad x'_\alpha = x_\alpha - \langle x_\alpha \rangle, \tag{14}$$

where $\langle x_\alpha \rangle = \frac{1}{S} \int x_\alpha dG$ is the coordinate of the center of mass in the initial system K, $\langle v_\alpha \rangle = \frac{d\langle x_\alpha \rangle}{dt}$. In the new system the surface force is the same as in the initial system $f'_\alpha = f_\alpha$, but the volume force transforms to $F'_\alpha = F_\alpha - \Phi_\alpha$ and total force is equal to zero: $\Phi'_\alpha = 0$. So, we eliminated the total force Φ_α using a noninertial center-of-mass reference system K'.

The total moment of force in the new system stays unchanged: $M' = M$. To eliminate the total moment of force M we switch from the system K' to the rotating reference system K'':

$$v''_\alpha \to v'_\alpha - e_{\alpha\beta} x'_\beta \Omega, \tag{15}$$

where Ω is the angular velocity of the rigid-body rotation

$$I \frac{d\Omega}{dt} = M, \tag{16}$$

where $I = \rho \int x'_\alpha x'_\alpha dG$ is the moment of inertia of our system. In the new system the surface force is the same as in the initial system $f''_\alpha = f'_\alpha$, but the volume force transforms to:

$$F''_\alpha \to F'_\alpha + \rho \left(e_{\alpha\beta} x'_\beta \dot{\Omega} + 2 e_{\alpha\beta} v'_\beta \Omega + \Omega^2 x'_\alpha \right), \tag{17}$$

and the total moment of force is equal zero: $M'' = 0$. In case of a small Reynolds number, the Coriolis force $2\rho e_{\alpha\beta} v'_\beta \Omega$ is small compared with the viscous force.

So in case of the total force Φ_α and total moment of force M not equal to zero we can eliminate them using the noninertial reference system with the rigid-body motion due to the force and moment of force.

3. Pressure calculation

Let χ_α and ψ be smooth fields in the region G related by

$$\frac{\partial \chi_\alpha}{\partial x_\beta} + \frac{\partial \chi_\beta}{\partial x_\alpha} = 2\psi \delta_{\alpha\beta}. \tag{18}$$

Multiplying the equation of motion (1) by χ_α, integrating over G, and using (2), (3), (18), we obtain

$$\int p\psi \, dG = -\frac{1}{2} \int f_\alpha \chi_\alpha d\Gamma \tag{19}$$

In the special case when $\psi = 1$ the expression (18) gives us $\chi_\alpha = x_\alpha$ and, according with (19),

$$\int p \, dG = -\frac{1}{2} \int f_\alpha x_\alpha d\Gamma \tag{20}$$

see (Landau & Lifshitz, 1986). In a general case, according with (18), ψ is an arbitrary harmonic function and $\chi = \chi_1 + i\chi_2$ is the analytical function associated with ψ as

$$d\chi = (\psi + i\omega)dz \tag{21}$$

where ω is a harmonic function conjugate to ψ.

The expressions (18) and (19) are basic in our theory. There is also an alternative way to derive them. The equations of motion (1), continuity (2) and the boundary conditions (3) can be obtained from the variation principle (Berdichevsky, 2009).

$$\delta\left[\frac{1}{4\mu}\int\left(p_{\alpha\beta}p_{\alpha\beta} - 2p^2\right)dG - \int f_\alpha v_\alpha d\Gamma\right] = 0 \tag{22}$$

or

$$\frac{1}{2\mu}\int\left(p_{\alpha\beta}\delta p_{\alpha\beta} - 2p\delta p\right)dG - \int f_\alpha \delta v_\alpha d\Gamma = 0 \tag{23}$$

Since (23) is valid for arbitrary variations of pressure δp and velocity δv_α we choose them such that $p_{\alpha\beta}$ is left unchanged:

$$\delta p_{\alpha\beta} = -\delta p \cdot \delta_{\alpha\beta} + \mu\left(\frac{\partial \delta v_\alpha}{\partial x_\beta} + \frac{\partial \delta v_\beta}{\partial x_\alpha}\right) = 0. \tag{24}$$

In this case (23) gives us

$$\frac{1}{\mu}\int p\delta p dG + \int f_\alpha \delta v_\alpha d\Gamma = 0. \tag{25}$$

We introduce the one-parameter family of variations $\delta v_\alpha = \frac{\chi_\alpha}{2\mu}\delta\varepsilon$, $\delta p = \psi\delta\varepsilon$. Then (24) and (25) take the form (18) and (19).

Suppose $x \in R^N$. Then it follows from (18) that

$$(N-2)\frac{\partial^2\psi}{\partial x_\alpha \partial x_\beta} = 0. \tag{26}$$

Therefore, in the three-dimensional case ψ is a linear function. Only in the two-dimensional case ψ can be an arbitrary harmonic function. Formulating in terms of (3.5), only in the two-dimensional space there exists a non-trivial system of pressure and velocity variations providing zero stress tensor variation.

The complete set of analytical functions ζ_k in the region G with the multiply connected boundary Γ consists of functions of the form z_k, $\left(z - z_m^0\right)^{-k}$, where z_m^0 are fixed points, each situated in one bubble. The complete set of harmonic functions ψ_k can be obtained in the form of $\text{Re}\,\zeta_k$ and $\text{Im}\,\zeta_k$.

According with (1), (2) the pressure p is a harmonic function. We present it in the form

$$p = \sum_k p_k \psi_k.$$ (27)

Using the expression (19) we obtain the algebraic system for coefficients p_k :

$$\sum_k \left(\int \psi_k \psi_n dG \right) p_k = -\frac{1}{2} \int f_\alpha \chi_{\alpha n} d\Gamma, \quad n = 0, 1, \dots$$ (28)

4. Velocity calculation

The stress tensor, expressed in terms of the Airy function φ,

$$p_{\alpha\beta} = \frac{\partial^2 \varphi}{\partial x_\alpha \partial x_\beta} - \frac{\partial^2 \varphi}{\partial x_\gamma \partial x_\gamma} \delta_{\alpha\beta},$$ (29)

satisfies the equation of motion (1) identically. The boundary conditions (3) take the form

$$e_{\alpha\beta}\tau_\gamma \frac{\partial^2 \varphi}{\partial x_\beta \partial x_\gamma} = -f_\alpha, \quad x \in \Gamma,$$ (30)

where τ_γ are the components of the unit tangential vector to the boundary, its direction being matched to the direction of circulation. Integrating (30) along the component boundary Γ_k from a fixed point to an arbitrary one we obtain

$$\frac{\partial \varphi}{\partial x_\alpha} = e_{\alpha\beta} \int f_\alpha d\Gamma_k, \quad x \in \Gamma_k.$$ (31)

Using (1), (29) and the explicit form of the stress tensor, we get

$$d \left(\frac{\partial \varphi}{\partial x_\alpha} \right) = 2\mu dv_\alpha + d\Phi_\alpha, \quad x \in G,$$ (32)

where

$$d(\Phi_1 + i\Phi_2) = (p + i\Omega), \quad \Omega = \mu \left(\frac{\partial v_1}{\partial x_2} - \frac{\partial v_2}{\partial x_1} \right),$$ (33)

Ω is a harmonic function conjugate to p ,

$$\frac{\partial \Phi_\alpha}{\partial x_\beta} + \frac{\partial \Phi_\beta}{\partial x_\alpha} = 2p\delta_{\alpha\beta}.$$ (34)

Therefore

$$\Phi_\alpha = \sum_n p_n \chi_{\alpha n},$$ (35)

where p_k are the coefficients of the pressure expansion (27). These coefficients are the solution of the system (28). According with (32) the velocity in the region G can be presented in the form

$$v_\alpha = \frac{1}{2\mu}\left(\frac{\partial \varphi}{\partial x_\alpha} - \Phi_\alpha\right), \quad x \in G. \tag{36}$$

The first term in the right-hand part of (36) is the potential part of velocity; the second term is the vortex part.
The gradient of the Airy function on the boundary was calculated in (31). Then we can calculate the velocity on the boundary as

$$v_\alpha = \frac{1}{2\mu}\left(e_{\alpha\beta}\int f_\alpha d\Gamma_k - \Phi_\alpha\right), \quad x \in \Gamma_k. \tag{37}$$

The expression (37) gives us the explicit presentation of the velocity on the boundary.

5. Limitations for the motion of the boundary

5.1 The rate of change of region perimeter

The strong limitation for the motion of the boundary is based on a general expression regarding the rate of change of perimeter L. To obtain this expression we use the fact (Dubrovin at al, 1984) that

$$\frac{d|\Gamma|}{dt} = \int v_\alpha n_\alpha H d\Gamma, \tag{38}$$

where $H = \dfrac{\partial n_\beta}{\partial x_\beta}$ is the mean curvature of the boundary. In the 2D case $|\Gamma|$ is the perimeter

of the region, and in the 3D case $|\Gamma|$ is the area of the boundary. We introduce the operator

of differentiation along the boundary $D_{\alpha\beta} = n_\alpha \dfrac{\partial}{\partial x_\beta} - n_\beta \dfrac{\partial}{\partial x_\alpha}$. Then we can write (38) in the

form

$$\frac{dL}{dt} = \int v_\alpha D_{\alpha\beta} n_\beta d\Gamma. \tag{39}$$

Using the identity

$$\int D_{\alpha\beta}\Lambda d\Gamma = 0, \tag{40}$$

where Λ is an arbitrary field which is continuous on the boundary, and also the equation of continuity (2) and the boundary conditions (3) we can write (39) in the final form

$$\frac{d|\Gamma|}{dt} = -\int \frac{p + f_\alpha n_\alpha}{2\mu} d\Gamma. \tag{41}$$

This expression is valid for any flow of incompressible Newtonian liquid (without Stokes approximation), generally speaking, with variable viscosity. We will use it for a 2D flow ($|\Gamma| = L$ is the perimeter of region), in case of constant viscosity:

$$\frac{dL}{dt} = -\frac{1}{2\mu}\int (p + f_\alpha n_\alpha)\, d\Gamma. \tag{42}$$

5.2 The dynamics of bubbles due to capillarity and air pressure

Let's take into account the capillary forces on the boundary, the external pressure p_0 and the pressure inside of the bubbles $p_k = p_b$, $k = 1,2,...,m$, equal in every bubble. Then the boundary force has the form

$$f_\alpha = -\sigma n_\alpha \frac{\partial n_\beta}{\partial x_\beta} - p_k n_\alpha, \quad x \in \Gamma_k, \tag{43}$$

where σ is the coefficient of surface tension. Using (42), (43) we get

$$\frac{dL}{dt} = -\frac{1}{2\mu}\left[\int p\, d\Gamma - (p_0 L_0 + p_b L_b) + 2\pi\sigma(m-1)\right], \tag{44}$$

where L_0 and L_b are the perimeter of external boundary and the total perimeter of the bubbles correspondingly.
Using (20) we obtain

$$\int p\, dG = p_0 S + (p_0 - p_b)S_b + \frac{\sigma}{2}L, \tag{45}$$

where S and S_b are the area of region and the total area of the bubbles.
For $\psi = p$, $\chi_\alpha = \Phi_\alpha$, the expressions (19), (34), (37) give us

$$\int p^2\, dG = \frac{\sigma}{2}\left(p_0 L_0 + p_b L_b + \int p\, d\gamma\right) + p_0^2 S + \left(p_0^2 - p_b^2\right)S_b - \\ -\mu(p_0 - p_b)\frac{dS_b}{dt}. \tag{46}$$

Using (44) - (46) and the inequality $\int p^2\, dG \geq \frac{1}{S}\left(\int p\, dG\right)^2$ we obtain the differential inequality

$$\mu\left[\sigma\frac{dL}{dt} + (p_0 - p_b)\frac{dS_b}{dt}\right] \leq \\ \leq -(p_0 - p_b)\left[\sigma L_b + (p_0 - p_b)S_b\right] - \\ -\frac{1}{S}\left[(p_0 - p_b)S_b + \frac{\sigma}{2}L\right]^2 - \pi\sigma^2(m-1). \tag{47}$$

This expression gives us the possibility to obtain the strict limitations for the motion of the free boundary in some special cases.

5.3 The influence of capillary forces only

In this case the inequality (47) may be simplified:

$$\frac{dL}{dt} \leq -\frac{\sigma}{2\mu}\left[\frac{L^2}{2S} + 2\pi(m-1)\right]. \tag{48}$$

where m is the number of bubbles. Let $L_\infty = 2\sqrt{\pi S}$ be the asymptotic value of the perimeter and let $\tau = \frac{\sigma t}{2\mu}\sqrt{\frac{\pi}{S}}$ be the dimensionless time. Then, according with (48), $L(\tau) \leq L_{up}(\tau)$,

$$L_{up} = \frac{L_0 + L_\infty th(\tau)}{L_\infty + L_0 th(\tau)}, \qquad m = 0,$$

$$L_{up} = \frac{L_0 L_\infty}{L_\infty + L_0\tau}(L_\infty + L_0\tau), \qquad m = 1, \tag{49}$$

$$L_{up} = \frac{\sqrt{m-1}\left(L_0 - L_\infty\sqrt{m-1}\,tg\left(\sqrt{m-1}\,\tau\right)\right)}{\sqrt{m-1} + \lambda_0 tg\left(\sqrt{m-1}\,\tau\right)}, m \geq 2.$$

where $L_{up}(\tau)$ is the upper limitation for time dependence of the perimeter - see Fig.2. The perimeter of system L lies in the interval $L_\infty \leq L \leq L_{up}(\tau)$.

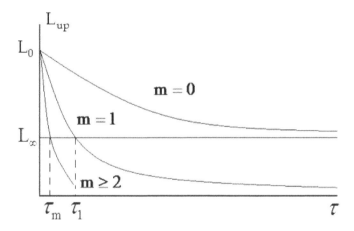

Fig. 2. The upper limitation for the time dependence of the perimeter for various number of bubbles m.

Therefore, if we have no bubbles in the region, the characteristic dimensionless time of relaxation of the boundary to the circle $\tau_0 \leq 1$. In case of one bubble $(m = 1)$, $L_{up}(\tau) \geq L_\infty$ at the time $\tau \leq \tau_1 = 1 - L_\infty/L_0$. The system with this topology can exist in this time period only. The bubble must collapse or break into two bubbles in time $\tau_* \leq \tau_1$. In case of $m > 2$ bubbles, such configuration will exist during the time

$$\tau \le \tau_m = \frac{1}{\sqrt{m-1}} arctg\left(\frac{\sqrt{m-1}\left(L_0 - L_\infty\right)}{L_0 + (m-1)L_\infty}\right). \tag{50}$$

5.4 Bubbles in an infinite region

The outer boundary of the region is a circle with a large radius R. The bubbles are localized around the center of the circle. Using the expressions $\pi R^2 - S_b = S$, $L = 2\pi R + L_b$, we can see that the inequality (47) in the limit $R \to \infty$ takes the form

$$\mu\frac{dW}{dt} \le -\left(p_0 - p_b\right)W - \pi\sigma^2 m , \tag{51}$$

where $W = \sigma L_b + (p_0 - p_b)S_b$. Therefore, at $p_0 - p_b > 0$

$$W + \frac{\pi\sigma^2 m}{p_0 - p_b} \le \left(W(0) + \frac{\pi\sigma^2 m}{p_0 - p_b}\right)\exp\left(-\frac{p_0 - p_b}{\mu}t\right). \tag{52}$$

Because $W \ge 0$, this configuration exists without change of the number of bubbles during the time

$$t \le \frac{\mu}{p_0 - p_b}\ln\left[1 + \frac{p_0 - p_b}{\pi\sigma^2 m}\left(\sigma L_b(0) + (p_0 - p_b)S_b(0)\right)\right]. \tag{53}$$

6. Motion of the boundary due to capillary forces

6.1 Calculation of pressure and velocity

In case of capillary forces action

$$f_\alpha = -\sigma n_\alpha\frac{\partial n_\beta}{\partial x_\beta}, \quad x \in \Gamma \tag{54}$$

and expression (19) takes the form

$$\int p\psi\,dG = \frac{\sigma}{2}\int \psi\,d\Gamma, \tag{55}$$

or

$$\langle p\psi\rangle_G = \langle p\rangle_G\langle\psi\rangle_\Gamma , \tag{56}$$

where

$$\langle f\rangle_G = \frac{1}{S}\int f dG, \quad \langle f\rangle_\Gamma = \frac{1}{L}\int f d\Gamma, \quad \langle P\rangle_G = \frac{\sigma}{2S}. \tag{57}$$

The expression (56) is valid for any harmonic function ψ. Let's apply $\psi = p$. Then we obtain

$$\langle p^2\rangle_G = \langle p\rangle_G\langle p\rangle_\Gamma , \tag{58}$$

It can be seen from (58) that

$$\langle p \rangle_\Gamma \ge \langle p \rangle_{\Gamma'}. \tag{59}$$

Introducing the generalized function (simple layer)

$$\delta_s(\mathbf{x}) = \int \delta(\mathbf{x} - \mathbf{y}) dl_{\mathbf{y}}, \tag{60}$$

we see that p is the projection of δ_s onto the subspace of harmonic functions.
Introducing in G a complete system of orthonormal harmonic functions $\{\Xi_k\}_{k=0}^{\infty}$ which obey the orthogonality condition $\langle \Xi_k \Xi_n \rangle_G = \delta_{kn}$, we obtain from (56) the following expression for the pressure

$$p = \langle p \rangle_G \sum_{k=0}^{\infty} \Xi_k \langle \Xi_k \rangle_\Gamma. \tag{61}$$

In case of capillary forces the expression (37) takes the form

$$v_\alpha = \frac{1}{2\mu}(\sigma n_\alpha - \Phi_\alpha), \quad x \in \Gamma. \tag{62}$$

6.2 Relaxation of a small perturbation of a circular cylinder

Consider a small perturbation of the circular cylinder boundary, given by $r = R + h(\varphi, t)$, $|h| \ll R$. Then we have from (62)

$$\frac{\partial h}{\partial t} = -\frac{\sigma}{2\mu R} \sum_{k=-\infty}^{\infty} |k| \exp(ik\varphi) h_k, \tag{63}$$

$$h_k(t) \equiv \int_0^{2\pi} \exp(-ik\varphi) h(\varphi, t) \frac{d\varphi}{2\pi} = h_k(0) \exp\left(-\frac{\sigma|k|t}{2\mu}\right), \tag{64}$$

in agreement with (Levich, 1962). According with (64), a small boundary perturbation of characteristic with $a \ll R$ and amplitude $H \ll a$ has a characteristic decay time $\tau \sim \dfrac{\mu a}{\sigma}$.

6.3 The capillary relaxation of an ellipse

Let's test our theory on an example of a large amplitude perturbation. We calculate the capillary relaxation of boundary with initial shape $\dfrac{x_1^2}{a^2} + \dfrac{x_2^2}{b^2} = 1$ in two ways - using the numerical calculation based on (6.4) and the finite-element software ANSYS POLYFLOW (see Fig. 3 and Fig.4). These methods of calculation give us the same results with discrepancy about 1%.

6.4 The collapse of a cavity

Let's now consider a large amplitude perturbation in the shape of a cavity (Fig. 5). By symmetry, the pressure must be an even function with respect to x_2, i.e. $p(x_1, -x_2) = p(x_1, x_2)$.

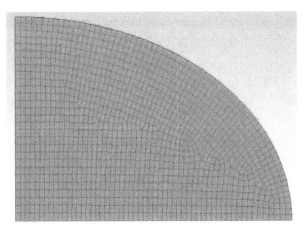

Fig. 3. Computational domain used in finite-element calculation of ellipse relaxation.

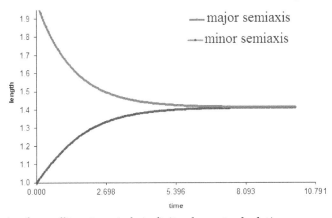

Fig. 4. Relaxation from ellipse to a circle in finite-element calculation.

We introduce a space of two-variable harmonic functions which are even with respect to the second argument, and choose in it the complete system of functions in the form $\psi_n = r^n \cos(n\varphi)$ (r and φ are the polar coordinates in the x_1, x_2 plane). Since the width δ is small $\langle \psi_m \psi_n \rangle_g = R^{2n} \dfrac{\delta_{mn}}{2(n+1)}$. Then the complete system of orthogonal harmonic functions in this space is

$$\Xi_n = \sqrt{2(n+1)}\left(\frac{r}{R}\right)^n \cos(n\varphi). \tag{65}$$

Inserting (65) in (61) and summing the series yields

$$p = \sigma\left[\frac{1}{R} - \frac{H}{\pi R^2} - \frac{2}{\pi}\operatorname{Re}\left(\frac{1}{R-z} - \frac{R-H}{R^2 - (R-H)R}\right)\right], \tag{66}$$

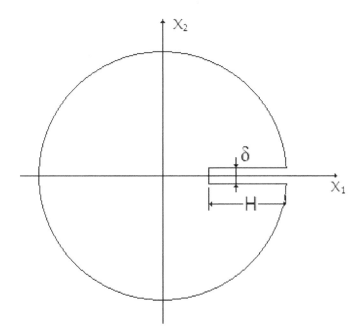

Fig. 5. Cavity perturbation.

whence, using (35), we have

$$\Phi = \sigma \left[\left(1 - \frac{H}{\pi R} \right) \frac{z}{R} + \frac{2}{\pi} \ln \left(\frac{R^2 - (R-H)z}{(R-H)R} \right) \right]. \tag{67}$$

In spite of the logarithm, (67) is a single-valued analytical function in G, because the boundary perturbation constitutes a branch cut. If we insert (67) in (62), we find that the normal velocity of the cut edges $V = \dfrac{\sigma}{2\mu}$ (in the zero approximation with respect to the small parameter $\dfrac{\delta}{H}$). The edges close up after a time $\tau = \dfrac{\mu\delta}{\sigma}$. Although capillary forces generally tend to flatten the boundary perturbation, in this case they produce the opposite effect. Acting to reduce the length of the cut, the capillary forces generate a flow of scale H in the region. The velocities along x_1 and x_2 have the scales $\dot{H}$ and $\dot{\delta}$, respectively. If we equate the work of surface-tension force with the rate of energy dissipation by viscous forces, we find that $\sigma\dot{H} \approx -\mu \left(\dfrac{\dot{H}}{H} \right)^2 H^2$ or $\dot{H} \approx \dot{\delta} \approx -\dfrac{\sigma}{\mu}$; this conforms to the rigorous result we obtained before.

7. Conclusion

We presented a method to calculate two-dimensional Stokes flow with free boundary, based on the expansion of pressure in a complete system of harmonic functions. The theory forms the basis for strict analytical results and numerical approximations. Using this approach we analyse the collapse of bubbles and relaxation of boundary perturbation. The results obtained by this method are correlating well with numerical calculations performed using commercial FEM software.

8. Acknowledgment

The authors would like to express their sincere gratitude to Prof. V. Pukhnachov and Prof. C. Pozrikidis for their attention to this research.

9. References

Antanovskii, L.K. (1988).Interface boundary dynamics under the action of capillary forces. Quasisteady-state plane-parallel motion., *J. Appl. Mech. and Techn. Phys.* Vol. 29, No 3, pp. 396-399.

Berdichevsky, V. (2009). *Variational principles of continuum mechanics*, Vol. 1, Springer-Verlag, Berlin-Heidelberg

Chivilikhin, S.A. (1992). Plane capillary flow of a viscous fluid with multiply connected boundary in the Stokes approximation, *Fluid Dynamics*, Vol. 27, No. 1, pp. 88 - 92.

Frenkel, J. (1945). Viscous flow of crystalline bodies under the action of surface tension, *Journal of. Physics*, Vol. 9, No. 5, pp. 385-391

Dubrovin, B.A., Fomenko, A.T. & Novikov, S.P. (1984). *Modern Geometry. Methods and Applications.* Part 1. Springer–Verlag

Happel, S.J. & Brenner, H. (1965). *Low Reynolds Number Hydrodynamics*, Prentice-Hall, Englewood Cliffs

Hopper, R.W. (1984). Coalescence of two equal cylinders: exact results for creeping viscous plane flow driven by capillarity, *J. Am. Ceram. Soc.*, Vol. 67, No. 12, pp. 262 - 264.

Ionesku, D.G. (1965). Theory of analytic functions and hydrodynamics, *in: Applications of the Theory of Functions in the Mechanics of Continuous Media*, Vol. 2, Mechanics of Liquids and Gases [in Russian], Nauka, Moscow.

Jeong, J.-T. & Moffatt, H.K. (1992). Free-surface cusps associated with flow at low Reynolds number, *J. Fluid Mech.*, Vol. 241, pp. 1-22.

Landau, L.D. & Lifshitz E.M. (1986). Theory of Elasticity. Course of Theoretical Physics. Vol.7. Butterworth-Heinemann.

Levich, V.G. (1962). *Physicochemical hydrodynamics*, Prentice-Hall, Englewood Cliffs, New Jersey

Muskeleshvili, N.I. (1966). *Some Fundamental Problems of the Mathematical Theory of Elasticity* [in Russian], Nauka, Moscow

Pozrikidis, C. (1997). Numerical studies of singularity formation at free surfaces and fluid interfaces in two-dimensional Stokes flow, *J. Fluid Mech.* 331, pp. 145-167.

Richardson, S. (2000). Plane Stokes flow with time-dependent free boundaries in which the fluid occupies a doubly-connected region, *Eur. J. Appl. Math.*, 11, pp. 249-269.

Tanveer, S. & Vasconcelos, G.L. (1994). Bubble Breakup in two-dimensional Stokes flow, *Phys. Rev. Lett.*, Vol. 73, No. 21, pp. 2845-2848.

3

Generalized Variational Principle for Dissipative Hydrodynamics: Shear Viscosity from Angular Momentum Relaxation in the Hydrodynamical Description of Continuum Mechanics

German A. Maximov
N. N. Andreyev Acoustical Institute
Russia

1. Introduction

A system of hydrodynamic equations for a viscous, heat conducting fluid is usually derived on the basis of the mass, the momentum and the energy conservation laws (Landau & Lifshitz, 1986). Certain assumptions about the form of the viscous stress tensor and the energy density flow vector are made to derive such a system of equations for the dissipative viscous, heat conductive fluid. The system of equations based on the mass, the momentum and the energy conservation laws describes adequately a large set of hydrodynamical phenomena. However, there are some aspects which suggest that this system is only an approximation.

For example, if we consider propagation of small perturbations described by this system, then it is possible to separate formally the longitudinal, shear and heat or entropy waves. The coupling of the longitudinal and heat waves results in their splitting into independent acoustic-thermal and thermo-acoustic modes. For these modes the limits of phase velocities tends to infinity at high frequencies so that the system is in formal contradiction with the requirements for a finite propagation velocity of any perturbation which the medium can undergo. Thus it is possible to suggest that such a hydrodynamic equation system is a mere low frequency approximation. Introducing the effects of viscosity relaxation (Landau & Litshitz, 1972), guarantees a limit for the propagation velocity of the shear mode, and the introduction of the heat relaxation term (Deresiewicz, 1957; Nettleton, 1960; Lykov, 1967) in turn ensures finite propagation velocities of the acoustic-thermal and thermo-acoustic modes. However, the introduction of such relaxation processes requires serious effort with motivation.

Classical mechanics provides us with the Lagrange's variational principle which allows us to derive rigorously the equations of motion for a mechanical system knowing the forms of kinetic and potential energies. The difference between these energies determines the form of the Lagrange function. This approach translates directly into continuum mechanics by introduction of the Lagrangian density for non-dissipative media. In this approach the dissipation forces can be accounted for by the introduction of the dissipation function derivatives into the corresponding equations of motion in accordance with Onsager's principle of symmetry of kinetic coefficients (Landau & Lifshitz, 1964). There is an established opinion that for a dissipative system it is impossible to formulate the variational principle analogously to the least action principle of Hamilton (Landau & Lifshitz, 1964). At the same time there are successful approaches (Onsager, 1931a, 1931b; Glensdorf & Prigogine, 1971; Biot, 1970; Gyarmati, 1970; Berdichevsky, 2009) in which the variational principles for heat conduction theory and for irreversible thermodynamics are applied to account explicitly for the dissipation processes. In spite of many attempts to formulate a variational principle for dissipative hydrodynamics or continuum mechanics (see for example (Onsager, 1931a, 1931b; Glensdorf & Prigogine, 1971; Biot, 1970; Gyarmati, 1970; Berdichevsky, 2009) and references inside) a consistent and predictive formulation is still absent. Therefore, there are good reasons to attempt to formulate the generalized Hamilton's variational principle for dissipative systems, which argues against its established opposition (Landau & Lifshitz, 1964). Thus the objective of the chapter is a new formulation of the generalized variational principle (GVP) for dissipative hydrodynamics (continuum mechanics) as a direct combination of Hamilton's and Osager's variational principles. The first part of the chapter is devoted to formulation of the GVP by itself with application to the well-known Navier-Stokes hydrodynamical system for heat conductive fluid. The second part of the chapter is devoted to the consistent introduction of viscous terms into the equation of fluid motion on the basis of the GVP. Two different approaches are considered. The first one dealt with iternal degree of freedom described in terms of some internal parameter in the framework of Mandelshtam – Leontovich approach (Mandelshtam & Leontovich, 1937). In the second approach the rotational degree of freedom as independent variable appears additionally to the mean mass displacement field. For the dissipationless case this approach leads to the well-known Cosserat continuum (Kunin, 1975; Novatsky, 1975; Erofeev, 1998). When dissipation prevails over angular inertion this approach describes local relaxation of angular momentum and corresponds to the sense of internal parameter. Finally, it is shown that the nature of viscosity phenomenon can be interpreted as relaxation of angular momentum of material points on the kinetic level.

2. Generalized variational principle for dissipative hydrodynamics

2.1 Hamilton's variational principle

The non-dissipative case of Hamilton's variational principle can be formulated for a continuous medium in the form of the extreme condition for the action functional $\delta S = 0$:

$$S = \int_{t_1}^{t_2} dt \int_V d\vec{r} L , \qquad (1)$$

which is an integral over the time interval (t_1 , t_2) and the initial volume V of a given mass of a continuum medium in terms of Lagrangian's coordinates. From the principles of particle mechanics the Lagrangian density L is represented as the difference between the kinetic K and potential U energies:

$$L(\dot{\vec{u}}, \nabla \vec{u}) = K(\dot{\vec{u}}) - U(\nabla \vec{u}) . \qquad (2)$$

Expression (2) implies that the Lagrangian can be considered as a function of the velocities of the displacements $\dot{\vec{u}} = \dfrac{\partial \vec{u}}{\partial t}$ and deformations $\nabla \vec{u} = div(\vec{u})$.

Generalized Variational Principle for Dissipative Hydrodynamics: Shear Viscosity from Angular Momentum Relaxation
in the Hydrodynamical Description of Continuum Mechanics

53

The motion equations derived from variational principles (1), (2) have the following form

$$\frac{d}{dt}\frac{\partial L}{\partial \dot{\vec{u}}} + \nabla \cdot \frac{\partial L}{\partial \nabla \vec{u}} = 0 .$$

(3)

In the simplest case, when the kinetic and potential energies are determined by the quadratic forms

$$2K(\dot{\vec{u}}^2) = \rho_0 \dot{\vec{u}}^2 , \quad 2U = \lambda \varepsilon_{ll}^2 + 2\mu \varepsilon_{ik}^2 , \quad \varepsilon_{ik} = \frac{1}{2}\left(\frac{\partial u_i}{\partial x_k} + \frac{\partial u_k}{\partial x_i}\right)$$

(4)

the well-known equation of motion for an elastic medium (Landau & Lifshitz, 1972) can be derived:

$$\rho_0 \frac{d}{dt}\dot{\vec{u}} - \mu \Delta \vec{u} - (\lambda + \mu)\nabla(\nabla \vec{u}) = 0 ,$$

(5)

where ρ_0 is the density of the medium, and λ and μ are the Lamé's constants.

2.2 Onsager's variational principle

If we consider quasi-equilibrium systems, then the Onsager's variational principle for least energy dissipation can be formulated (Onsager, 1931a, 1931b). This principle is based on the symmetry of the kinetic coefficients and can be formulated as the extreme condition for the functional constructed as the difference between the rate of increase of entropy, $\dot{s}$, and the dissipation function, D. Here the entropy s is considered as a function of some thermodynamic relaxation process α, and the dissipation function D as a function of the rate of change of α, i.e.

$$\delta_{\dot{\alpha}}[\dot{s}(\alpha) - D(\dot{\alpha})] = 0$$

(6)

The kinetic equation can then be derived from variational principle (6) to describe the relaxation of a thermodynamic system to its equilibrium state, i.e.:

$$\frac{d}{dt}s(\alpha) = 2D(\dot{\alpha}) .$$

(7)

The above equation satisfies strictly the symmetry principle for the kinetic coefficients (Landau & Lifshitz, 1986).

2.3 Variational principle for mechanical systems with dissipation

As was mentioned above, the generalization of the equation of motion (3) in the presence of dissipation is obtained by introducing the derivative of the dissipation function with respect to the velocities into the right hand side of the equation (3). Therefore, in accordance with Onsager's symmetry principle for the kinetic coefficients (Landau & Lifshitz, 1964) we have

$$\frac{d}{dt}\frac{\partial L}{\partial \dot{\vec{u}}} + \nabla \cdot \frac{\partial L}{\partial \nabla \vec{u}} = -\frac{\partial D}{\partial \dot{\vec{u}}} .$$

(8)

Now it is possible to show, that the equation of motion can be derived in the form of equation (8) if Hamilton's variational principle is adapted with the following form of the Lagrangian function:

$$L(\dot{\vec{u}},\nabla\vec{u}) = K(\dot{\vec{u}}) - U(\nabla\vec{u}) - \int_0^t D(\dot{\vec{u}})dt' ,\qquad(9)$$

where the time integral of the dissipation function is introduced into equation (2). The initial time in integral (9) denoted for simplicity equal to 0 corresponds to the time t_1 in functional (1).

It needs, however, to pay attention that at variation of dissipative term in such approach an additional item appears, which has to be neglected by hands. Indeed, variation of the last term in (9) leads us to result

$$\delta\int_0^t D(\dot{\vec{u}})dt' = \int_0^t \frac{\partial D(\dot{\vec{u}})}{\partial\dot{\vec{u}}}\,\delta\dot{\vec{u}}\,dt' = \int_0^t \frac{d}{dt'}\left(\frac{\partial D(\dot{\vec{u}})}{\partial\dot{\vec{u}}}\,\delta\vec{u}\right)dt' - \int_0^t \frac{d}{dt'}\left(\frac{\partial D(\dot{\vec{u}})}{\partial\dot{\vec{u}}}\right)\delta\vec{u}dt' \qquad(10a)$$

If to neglect by the last item in this expression

$$\delta\int_0^t D(\dot{\vec{u}}(t'))dt' = \frac{\partial D(\dot{\vec{u}})}{\partial\dot{\vec{u}}}\,\delta\vec{u}(t) - \int_0^t \frac{d}{dt'}\left(\frac{\partial D(\dot{\vec{u}})}{\partial\dot{\vec{u}}}\right)\delta\vec{u}dt' \approx \frac{\partial D(\dot{\vec{u}})}{\partial\dot{\vec{u}}}\,\delta\vec{u}(t) ,\qquad(10b)$$

then the result gives us the same term $\dfrac{\partial D(\dot{\vec{u}})}{\partial\dot{\vec{u}}}$, which we need artificially introduce in the motion equation (8) for account of dissipation. From the one hand this approach can be considered as some rule at variation of integral term, because it leads us to the required form of the motion equation (8). From the other hand the following supporting basement can be proposed. Variation of action containing all terms in Lagrangian (9) with account of initial and boundary conditions can be written in the form

$$\int_{t_1}^{t_2} dt\int dV\left\{\left(-\frac{d}{dt}\frac{\partial K(\dot{\vec{u}})}{\partial\dot{\vec{u}}} + \nabla\frac{\partial U(\nabla\vec{u})}{\partial\nabla\vec{u}} - \frac{\partial D(\dot{\vec{u}})}{\partial\dot{\vec{u}}}\right)\delta\vec{u} + \int_0^t \frac{d}{dt'}\left(\frac{\partial D(\dot{\vec{u}})}{\partial\dot{\vec{u}}}\right)\delta\vec{u}dt'\right\} =\qquad(11a)$$

It is seen from (11a) that the required form of the motion equation with dissipation arises due to zero value of coefficient at arbitrary variation of the displacement field $\delta\vec{u}$. The last additional item, containing variation $\delta\vec{u}$ under integral, prevent to the strict conclusion in the given case. Nevertheless, if to rewrite the first term in (11a) in the same integral form as the last term

$$= \int_{t_1}^{t_2} dt\int dV\int_0^t dt'\left\{\delta(t-t')\left(-\frac{d}{dt'}\frac{\partial K(\dot{\vec{u}})}{\partial\dot{\vec{u}}} + \nabla\frac{\partial U(\nabla\vec{u})}{\partial\nabla\vec{u}} - \frac{\partial D(\dot{\vec{u}})}{\partial\dot{\vec{u}}}\right) + \frac{d}{dt'}\left(\frac{\partial D(\dot{\vec{u}})}{\partial\dot{\vec{u}}}\right)\right\}\delta\vec{u}\qquad(11b)$$

then due to the same reason of arbitrary variation $\delta\vec{u}$ the multiplier in brackets at this variation has to be equal to zero. It is possible to see now, that, if the function $\dfrac{d}{dt'}\left(\dfrac{\partial D(\dot{\vec{u}})}{\partial\dot{\vec{u}}}\right)$ is not singular in the point $t' = t$, then its contribution can be neglected in this point in comparison with singular contribution from delta-function. The presented arguments can be considered as a basis for variation rule of integral term in Lagrangian.

In particular, if the dissipation function is considered as a quadratic form of the deformation velocities, i.e.:

$$2D(\nabla \dot{\vec{u}}) = \eta' \left(\frac{\partial \dot{u}_i}{\partial x_k} + \frac{\partial \dot{u}_k}{\partial x_i} \right)^2 + \varsigma' \left(\frac{\partial \dot{u}_l}{\partial x_l} \right)^2 , \tag{12}$$

then the derived equation of motion with account of (4) corresponds to the linearized Navier–Stokes equation:

$$\rho_0 \frac{d}{dt} \dot{\vec{u}} - (\lambda + \mu)\Delta\vec{u} - \lambda\nabla(\nabla\vec{u}) = \eta\Delta\dot{\vec{u}} + \left(\varsigma + \frac{\eta}{3} \right)\nabla(\nabla\dot{\vec{u}}) , \tag{13}$$

where the shear and volume viscosities, η and ς respectively are given by $\eta'/2$ and $\varsigma' + \frac{4}{3}\eta'$ respectively, from the constants in (12).

2.4 Independent variables

When GVP is formulated in the form (9) we need to determine variables in which terms the Lagrange's function has to be expressed. To answer on this question let's return to the hydrodinamics equations and look at variables for their description.

In absence of dissipation, as it easy to see, these variables are velocity, density, pressure and entropy $\vec{v}, \rho, P, s$. For the dissipationless case the entropy holds to be constant for given material point, hence a pressure can be considered, for example, as a function of solely density $P(\rho, s = const)$. The density of the given mass of continuum is expressed in terms of its volume. Hence variation of density can be expressed in terms of variation of volume or through divergence of the displacement field $\rho = \rho(div\vec{u})$. In particular, linearization of the continuity equation leads to relation

$$\rho = \rho_0(1 - div\vec{u}) \tag{14}$$

Velocity by definition is a time derivative from displacement $\vec{v} = \dot{\vec{u}}$. Thus, the displacement field $\vec{u}$ can be considered as the principal hydrodinamical variable for the dissipationless case.

In the presence of dissipation, the hydrodynamic equations also involve the temperature T, implying in the following set of variables: $\vec{v}, \rho, P, s, T$. If pressure and entropy depend on density and temperature $P(\rho, T)$, $s(\rho, T)$ in accordance to the state equation, then the fields of displacements and temperatures: $\vec{u}, T$ can be considered as the principal hydrodynamical variables.

Further, we will adopt the idea of Biot (Biot, 1970), and introduce some vector field $\vec{u}_T$ (some vector potential), called the heat displacement, as independent variable instead temperature, so that the relative deviation of temperature T from its equilibrium state T_0 is determined by the divergence of the field $\vec{u}_T$. Namely in analogy with (14)

$$T = T_0(1 - \theta \, div\vec{u}_T) \tag{15a}$$

where θ is some dimensionless constant which is specially introduced in definition (15a) for simplification of the expression for the dissipation function. Thus, the divergence of the heat displacement field $\vec{u}_T$ determines temperature deviation from its equilibrium level

$$\frac{T - T_0}{T_0} = -\theta \nabla \vec{u}_T .$$

(15b)

2.5 Generalized variational principle (GVP) for dissipative hydrodynamics

The above example (12), (13) of the derivation of the equation of motion for dissipative systems on the basis of Hamilton's variational principle with the Lagrange's function (9) suggests the possibility of formulating a generalized variational principle for dissipative hydrodynamical systems. This formulation can be obtained by a simple combination of Hamilton's variational principle (eqs. (1) and (2)) and Onsager's variational principle (eq. (6)), if the latter is integrated over time and multiplied by a temperature term (Maximov , 2008, 2010, originally formulated by Maximov , 2006). The Lagrangian density in this case can be written in the following form:

$$L = K - E + T\left[s - \int_0^t D dt' \right] = K - F - T\int_0^t D dt' ,$$

(16)

where E and F are the internal energy (potential for the dissipationless case) and the free energy respectively. For the non-dissipative case, the Lagrangian depends on the time and spatial derivatives of the mean mass displacement field $\vec{u}$, which is a basic independent variable in this formulation. For the dissipative case, the temperature should be considered as an additional independent variable for a complete description. Hence, a free energy and dissipation function should also depend on the temperature variations. But temperature by itself is not a convenient variable here. Instead it is more convenient to consider the heat displacements $\vec{u}_T$, introduced in previous section, of which the divergence will give us temperature.

In this case the generalized Lagrangian can be written in the following form:

$$L(\dot{\vec{u}}, \nabla \vec{u}, \nabla \vec{u}_T) = K(\dot{\vec{u}}) - F(\nabla \vec{u}, \nabla \vec{u}_T) - T_0 \int_0^t D(\dot{\vec{u}}, \dot{\vec{u}}_T) dt' .$$

(17)

It is important to note here that the opportunity to formulate the variational principle for a dissipative system arises due to the energy conservation for two interacting fields: the mean mass displacement $\vec{u}$ and the heat displacement $\vec{u}_T$. The dissipation function only plays a role in the transformation rate between these fields.

In this way the motion equations derived by variation of action with the Lagrangian (17), can be expressed in the following forms

$$\frac{d}{dt}\frac{\partial K}{\partial \dot{\vec{u}}} - \nabla \frac{\partial F}{\partial \nabla \vec{u}} = -T_0 \frac{\partial D}{\partial \dot{\vec{u}}} ,$$

$$T_0 \frac{\partial D}{\partial \dot{\vec{u}}_T} - \nabla \frac{\partial F}{\partial \nabla \vec{u}_T} = 0 .$$

(18)

Taking into account that the kinetic energy is given by quadratic form (4), the free energy is given by its usual expression for thermo-elasticity in quadratic form (Landau & Lifshitz, 1972):

$$2F(\nabla \vec{u}, T) = 2\mu \varepsilon_{ik}^2 + \lambda \varepsilon_{ll}^2 + \tilde{\kappa} \left(\frac{T - T_0}{\theta T_0} \right)^2 + 2\tilde{\alpha} \varepsilon_{ll} \left(\frac{T - T_0}{\theta T_0} \right), \tag{19a}$$

or with substitution of expression (13) instead of the temperature terms:

$$2F(\nabla \vec{u}, \nabla \vec{u}_T) = 2\mu \varepsilon_{ik}^2 + \lambda \varepsilon_{ll}^2 + \tilde{\kappa} (\nabla \vec{u}_T)^2 + 2\tilde{\alpha} \varepsilon_{ll} (\nabla \vec{u}_T) \tag{19b}$$

The dissipation function is the square of the difference between the mean mass and the heat displacements

$$2D(\dot{\vec{u}}, \dot{\vec{u}}_T) = \beta (\dot{\vec{u}} - \dot{\vec{u}}_T)^2 \tag{20}$$

The meanings of the coefficients $\tilde{\kappa}$, $\tilde{\alpha}$ and β in quadratic forms (19), (20) will be defined in the next section by comparison with the classical Navier-Stokes hydrodynamical system of equations.

In this case the motion equations for the mean displacement field and for the temperature field derived on the basis of the generalized variational principle are just equivalent (at $\mu = 0$) to the linearized traditional system of hydrodynamics equations:

$$\rho_0 \frac{d}{dt} \dot{\vec{u}} - \mu \Delta \vec{u} - (\lambda + \mu + \tilde{\alpha}) \nabla (\nabla \vec{u}) = (\tilde{\alpha} + \tilde{\kappa}) / (\theta T_0) \nabla T \tag{21}$$

$$\beta (\dot{T} - T_0 \theta \nabla \dot{\vec{u}}) - \tilde{\kappa} \Delta T = \tilde{\alpha} T_0 \theta \Delta \nabla \vec{u}. \tag{22}$$

2.6 Comparison with the system of hydrodynamics equations

Coefficients of the quadratic forms in equations (19) and (20) can be determined by comparison between the system of equation (21) and (22) and the linearized system of hydrodynamics equations (Landau & Lifshitz, 1986) considering the variables $\vec{u}, T$:

$$\rho = \rho_0 (1 - \nabla \vec{u}), \tag{23}$$

$$\rho_0 \frac{d^2 \vec{u}}{dt^2} - \rho_0 c_0^2 \Delta \vec{u} = -\rho_0 \alpha \nabla T + \eta \Delta \dot{\vec{u}} + \left(\zeta + \frac{\eta}{3} \right) \nabla (\nabla \dot{\vec{u}}), \tag{24}$$

$$\rho_0 C_V \frac{dT}{dt} + \rho_0 T_0 \alpha \nabla \dot{\vec{u}} - \kappa \Delta T' = 0. \tag{25}$$

where c_0 is the isothermal sound velocity, C_V - the heat capacity at constant volume, κ the heat conductivity coefficient, and α the thermal expansion coefficient. In the absence of viscosity $\eta = 0$ and $\varsigma = 0$, which was not taken into account in the dissipation function (20), the structure of equations (21), (22) nearly coincides with the second (24) and the third (25) equations in the system of hydrodynamics equations (Landau & Lifshitz, 1986). The only difference is the additional term in the right part of equation (22) in comparison with (25). We note here briefly that the reason for the introduction of this term is related to a generalized form of the Fourier law for heat energy flow. Besides the term of the temperature gradient in the Fourier law, an additional density or pressure gradient term should appear in spite of the contradicting argument presented in (Landau & Lifshitz, 1986). The independent support of this result can be found in refs. (Martynov, 2001; Zhdanov & Roldugin 1998).

The coefficients of equations (21), (22) and (24), (25) for the fluid case ($rot(\vec{u}) = 0$) can be found by comparison. One needs to take into account the different dimensions of equation (22) and (25), and, hence, the presence of common dimension multiplier in the comparison of coefficients for these equations.

The parameters of the quadratic forms are expressed explicitly in terms of the physical parameters by the following expressions

$$\beta = \frac{\rho_0 c_0^2}{\chi}\left(\gamma^2 - 1\right), \ \theta = -\frac{\gamma - 1}{\alpha T_0}, \ \tilde{\alpha} = \rho_0 c_0^2\left(\gamma - 1\right), \ \lambda + 2\mu = \rho_0 c_0^2 \gamma, \ \tilde{\kappa} = \rho_0 c_0^2\left(\gamma^2 - 1\right), \quad (26)$$

where γ is the specific temperature ratio, $\gamma = C_P / C_V$, and $\chi = \kappa / \rho_0 C_V$ is the temperature conductivity coefficient. It is remarkable that the coefficient in the dissipation function β is inversely proportional to the temperature conductivity coefficient.

3. Viscous terms in dissipative hydrodynamics

3.1 Account of viscosity relaxation for a fluid

To take into account fluid viscosity in the equation of motion in the framework of the generalized variational principle it is possible to introduce additional internal parameters to describe the quasi-equilibrium state of the medium, analogous to the Mandelshtam – Leontovich approach (Mandelshtam & Leontovich, 1937). As will be shown, in order to describe both the shear and the volume viscosities simultaneously, this internal parameter needs to possess the properties of a tensor. To simplify the description we consider the case when the temperature variation variable T is not essential so that the heat displacement $\vec{u}_T$ terms can be omitted. In this case the additional terms associated with the tensor internal parameter ξ_{ik}, will appear in the expression for the free energy of an elastic medium (19), and it can be written as:

$$2F(\nabla \vec{u}, \ \xi_{ij}) = 2\mu\varepsilon_{ik}^2 + \lambda\varepsilon_{ll}^2 + a_1\xi_{ll}^2 + a_2\xi_{ik}^2 + 2b_1\xi_{kk}\varepsilon_{ll} + 2b_2\xi_{ik}\varepsilon_{ki}, \qquad (19c)$$

where a_i and b_i are some coefficients of a positively determined quadratic form. The kinetic energy is then given by the ordinal expression (4) and the dissipative function in the absence of the temperature term can be written as the following quadratic form:

$$2D(\dot{\xi}_{ij}) = \gamma_1\dot{\xi}_{ll}^2 + \gamma_2\dot{\xi}_{ik}^2 \qquad (27)$$

with some coefficients γ_1, γ_2.

The system of motion equations, derived on the basis of the generalized variational principle for this case can be rewritten as

$$\rho_0 \frac{d}{dt}\dot{\bar{u}} - \mu\Delta\bar{u} - (\lambda+\mu)\nabla(\nabla\bar{u}) - b_1\nabla\xi_{ll} - b_2\frac{\partial\xi_{ik}}{\partial x_k} = 0, \tag{28}$$

$$\gamma_1\delta_{ik}\frac{d\xi_{ll}}{dt} + \gamma_2\frac{d\xi_{ik}}{dt} + a_1\delta_{ik}\xi_{ll} + a_2\xi_{ik} + b_1\delta_{ik}\nabla\bar{u} + b_2\varepsilon_{ik} = 0. \tag{29}$$

Here in the first equation (28) we safe for shortness the tensor notation for vector obtained as divergence of internal parameter tensor. Equation (28) is the motion equation for an elastic medium. Equation (29) is the kinetic equation for the internal parameter tensor ξ_{ik}. Convolving the kinetic equation by indexes it is possible to obtain the separate kinetic equation for the spherical part of the internal parameter tensor ξ_{ll}:

$$\tilde{\gamma}\frac{d\xi_{ll}}{dt} + \tilde{a}\xi_{ll} + \tilde{b}\varepsilon_{ll} = 0, \tag{30}$$

where the coefficients with tilde have the following meaning:

$$\tilde{\gamma} = 3\gamma_1 + \gamma_2, \quad \tilde{a} = 3a_1 + a_2, \quad \tilde{b} = 3b_1 + b_2 \tag{31}$$

Kinetic equation (29) is an inhomogeneous ordinary differential equation of the first order. Its solution can be written as:

$$\xi_{ll} = -\frac{\tilde{b}}{\tilde{\gamma}}\int_{-\infty}^{t} e^{-\frac{\tilde{a}}{\tilde{\gamma}}(t-t')}\varepsilon_{ll}(t')\,dt' \tag{32}$$

For the other components of the internal tensor parameter ξ_{ik} we can also obtain a kinetic equation of similar form to equation (29), but with added inhomogeneous terms, i.e.

$$\gamma_2\frac{d\xi_{ik}}{dt} + a_2\xi_{ik} + b_2\varepsilon_{ik} + \tilde{a}_1\delta_{ik}\xi_{ll} + \tilde{b}_1\delta_{ik}\varepsilon_{ll} = 0 \tag{33}$$

where the following notations are introduced

$$\tilde{a}_1 = \left(a_1 - \tilde{a}\frac{\gamma_1}{\tilde{\gamma}}\right),$$

$$\tilde{b}_1 = \left(b_1 - \tilde{b}\frac{\gamma_1}{\tilde{\gamma}}\right) \tag{34}$$

Again, the solution of equation (33) has a form analogous to expression (32) with additional contributions from the terms with multipliers $\tilde{a}_1$ and $\tilde{b}_1$. Specifically,

$$\xi_{ik} = -\frac{b_2}{\gamma_2}\int_{-\infty}^{t} dt' e^{-\frac{a_2}{\gamma_2}(t-t')}\left(\varepsilon_{ik} - \delta_{ik}\varepsilon_{ll}\left(1 - \frac{\tilde{b}}{b_2}\frac{(a_1\gamma_2 - a_2\gamma_1)}{(\tilde{a}\gamma_2 - a_2\tilde{\gamma})}\right)\right) - \delta_{ik}\frac{\tilde{b}}{\tilde{\gamma}}\frac{(a_1\tilde{\gamma} - \tilde{a}\gamma_1)}{(\tilde{a}\gamma_2 - a_2\tilde{\gamma})}\int_{-\infty}^{t} dt' e^{-\frac{\tilde{a}}{\tilde{\gamma}}(t-t')}\varepsilon_{ll} \tag{35}$$

Taking the divergence of tensor (35), we obtain the following vector

$$\frac{\partial \xi_{ik}}{\partial x_k} = -\frac{b_2}{\gamma_2} \int_{-\infty}^{t} dt' e^{-\frac{a_2}{\gamma_2}(t-t')} \left(\frac{1}{2}\left(\Delta\vec{u} + \nabla(\nabla\vec{u})\right) - \nabla(\nabla\vec{u})\left(1 - \frac{\tilde{b}}{b_2}\frac{(a_1\gamma_2 - a_2\gamma_1)}{(\tilde{a}\gamma_2 - a_2\tilde{\gamma})}\right) \right) -$$

(36)

$$-\frac{\tilde{b}}{\tilde{\gamma}}\frac{(a_1\tilde{\gamma} - \tilde{a}\gamma_1)}{(\tilde{a}\gamma_2 - a_2\tilde{\gamma})} \int_{-\infty}^{t} dt' e^{-\frac{\tilde{a}}{\tilde{\gamma}}(t-t')} \nabla(\nabla\vec{u})$$

If we substitute (36) and (32) in the motion equation (28), we can write:

$$\rho_0 \frac{d}{dt}\dot{\vec{u}} - \mu\Delta\vec{u} - (\lambda + \mu)\nabla(\nabla\vec{u}) = -\frac{\tilde{b}}{\tilde{\gamma}}\left(b_1 - b_2\frac{(a_1\tilde{\gamma} - \tilde{a}\gamma_1)}{(\tilde{a}\gamma_2 - a_2\tilde{\gamma})}\right) \int_{-\infty}^{t} dt' e^{-\frac{\tilde{a}}{\tilde{\gamma}}(t-t')} \nabla(\nabla\vec{u}) -$$

(37)

$$-\frac{b_2^2}{\gamma_2} \int_{-\infty}^{t} dt' e^{-\frac{a_2}{\gamma_2}(t-t')} \left(\frac{1}{2}\left(\Delta\vec{u} + \nabla(\nabla\vec{u})\right) - \nabla(\nabla\vec{u})\left(1 - \frac{\tilde{b}}{b_2}\frac{(a_1\gamma_2 - a_2\gamma_1)}{(\tilde{a}\gamma_2 - a_2\tilde{\gamma})}\right) \right)$$

In the low frequency limit, at times greater than the relaxation times $\tilde{\gamma}/\tilde{a}$ and γ_2/a_2, it is possible to derive an equation analogous to the Navier – Stokes motion equation with shear and volume viscosities:

$$\rho_0 \frac{d}{dt}\dot{\vec{u}} - \tilde{\mu}\Delta\vec{u} - (\tilde{\lambda} + \tilde{\mu})\nabla(\nabla\vec{u}) = \tilde{\eta}\Delta\dot{\vec{u}} + (\tilde{\zeta} + \frac{\tilde{\eta}}{3})\nabla(\nabla\dot{\vec{u}})$$ (38)

where the effective elastic moduli $\tilde{\lambda}$ and $\tilde{\mu}$ and coefficients of shear and volume viscosities are expressed as

$$\tilde{\mu} = \mu - \frac{b_2^2}{2a_2}, \quad \tilde{\lambda} = \lambda + \frac{b_2^2}{2a_2} - \frac{\tilde{b}}{\tilde{a}}\left(b_1 - b_2\frac{(a_1\tilde{\gamma} - \tilde{a}\gamma_1)}{(\tilde{a}\gamma_2 - a_2\tilde{\gamma})}\right), \quad \tilde{\eta} = \frac{1}{2}\gamma_2\frac{b_2^2}{a_2^2},$$

(39)

$$\tilde{\zeta} + \frac{\tilde{\eta}}{3} = \tilde{\gamma}\frac{\tilde{b}}{\tilde{a}^2}\left(b_1 - b_2\frac{(a_1\tilde{\gamma} - \tilde{a}\gamma_1)}{(\tilde{a}\gamma_2 - a_2\tilde{\gamma})}\right) - \gamma_2\frac{b_2}{a_2^2}\left(\frac{b_2}{2} - \tilde{b}\frac{(a_1\gamma_2 - a_2\gamma_1)}{(\tilde{a}\gamma_2 - a_2\tilde{\gamma})}\right)$$

It is important to note that the structure of the effective shear modulus $\tilde{\mu}$ in (39) is determined by a difference, which can be equal to zero, in which case equation (38) becomes completely equivalent to the Navier – Stokes equation for a viscous fluid. Thus the condition

$$\mu = \frac{b_2^2}{2a_2}$$ (40)

should be satisfied to consider a solid with shear relaxation like a viscous fluid. If $\tilde{\mu} > 0$, then we have the case of elastic medium with a shear viscosity (the Voight's model) or with relaxation in the more general case (37). Thus, in the framework of the uniform approach it is possible to describe viscous fluids and solids with visco-elastic properties.

As a final remark of this section it is possible to say several words about physical sense of the introduced internal parameter. Since in the low frequency limit the majority of gases and fluids, including the simplest of them, is described by the Navier-Stokes equation, then the only available value, which could relax in all cases, and hence could be considered as common scalar internal parameter, is the mean distance between molecules in gas or liquid. In the condensed and especially in the solid media the mutual space placement of atoms becomes to be essential, hence a space variation of their mutual positions, holding rotational invariance of a body as whole, has to be described by symmetrical tensor of the second order. Hence the corresponding internal parameter could be the same tensor. Thus, the discrete structure of medium on the kinetic level predetermines existence, at least, of mentioned internal parameters, responsible for relaxation.

3.2 Shear viscosity as a consequence of the angular momentum relaxation for the hydrodynamical description of continuum mechanics

As shown in the previous section, it is possible to derive the system of hydrodynamical equations on the GVP basis for viscous, compressible fluid in the form of Navier-Stokes equations. However for the account of terms responsible for viscosity it is required to introduce some tensor internal parameter ξ_{ik} in agreement with Mandelshtam-Leontovich approach (Mandelshtam & Leontovich, 1937). Relaxation of this internal parameter provides appearance of viscous terms in the Navier-Stokes equation. It is worth mentioning that the developed approach allowed to generalize the Navier-Stokes equation with constant viscosity coefficient to more general case accounting for viscosity relaxation in analogy to the Maxwell's model (Landau & Lifshitz, 1972). However the physical interpretation of the tensor internal parameter, which should be enough universal due to general character of the Navier-Stokes equation, requires more clear understanding. On the intuition level it is clear that corresponding internal parameter should be related with neighbor order in atoms and molecules placement and their relaxation. In the present section such physical interpretation is represented.

As was mentioned in Introduction the system of hydrodynamical equations in the form of Navier-Stokes is usually derived on the basis of conservation laws of mass M, momentum $\vec{P}$ and energy E. The correctness of equations of the traditional hydrodynamics is confirmed by the large number of experiments where it is adequate. However the conservation law of angular momentum $\vec{M}$ is absent among the mentioned balance laws laying in the basis of traditional hydrodynamics. In this connection it is interesting to understand the role of conservation law of angular momentum $\vec{M}$ in hydrodynamical description. It is worth mentioning that equation for angular momentum appeared in hydrodynamics early (Sorokin, 1943; Shliomis, 1966) and arises and develops in the momentum elasticity theory. The Cosserat continuum is an example of such description (Kunin, 1975; Novatsky, 1975; Erofeev, 1998). However some internal microstructure of medium is required for application of such approach.

In the hydrodynamical description as a partial case of continuum mechanics the definition of material point is introduced as sifficiently large ensemble of structural elements of medium (atoms and molecules) that on one hand one has to describe properties of this ensemble in statistical way and on the other one has to consider the size of material point as small in comparison with specific scales of the problem. A material point itself as closed ensemble of particles possesses the following integrals of motion: mass, momentum, energy and angular momentum.

The basic independent variables, in terms of which the hydrodynamical description should be constructed, are the values which can be determined for separate material point in accordance with its integrals of motion: mean mass displacement vector $\vec{u}$ (velocity of this displacement $\vec{v} = \partial \vec{u} / \partial t$ is determined by integrals of motion $\vec{v} = \vec{P} / M$), rotation angle $\vec{\varphi}$ (angular velocity of rotation $\vec{\Omega} = \dot{\vec{\varphi}}$ is determined by integrals of motion $\vec{\Omega} = \vec{M} / I$, where I - inertia moment) and heat displacement $\vec{u}_T$, determining variation of temperature and related with integral of energy E.

In accordance with the set of independent field variables we can represent the kinetic K and the free F energies as corresponding quadratic forms

$$2K = \rho \dot{\vec{u}}^2 + I \dot{\vec{\varphi}}^2 \tag{41}$$

$$2F = (\lambda + 2\mu)(\nabla \vec{u})^2 + \mu[\nabla \vec{u}]^2 + 2\delta \vec{\varphi}[\nabla \vec{u}] + \sigma(\vec{\varphi})^2 + \varepsilon(\nabla \vec{\varphi})^2 + \varsigma[\nabla \vec{\varphi}]^2 \tag{42}$$

Taking into account that the dissipation dealt only with field of micro rotations, and omitting for shortness dissipation of mean displacement field, described by heat conductivity, we can write the dissipation function in the following form

$$2D = \gamma \dot{\vec{\varphi}}^2 \tag{43}$$

Equations of motion derived from GVP without temperature terms have the forms:

$$\frac{d}{dt}\frac{\partial K}{\partial \dot{\vec{u}}} - \nabla \frac{\partial F}{\partial \nabla \vec{u}} - [\nabla \frac{\partial F}{\partial [\nabla \vec{u}]}] = -\frac{\partial D}{\partial \dot{\vec{u}}} \tag{44a}$$

$$\frac{d}{dt}\frac{\partial K}{\partial \dot{\vec{\varphi}}} + \frac{\partial K}{\partial \vec{\varphi}} - \nabla \frac{\partial F}{\partial \nabla \vec{\varphi}} - [\nabla \frac{\partial F}{\partial [\nabla \vec{\varphi}]}] = -\frac{\partial D}{\partial \dot{\vec{\varphi}}} \tag{45a}$$

Without dissipation $\beta = 0$ the motion equations obtained with use of quadratic forms (41)-(43) correspond to the ones for Cosserat continuum (Kunin, 1975; Novatsky, 1975; Erofeev, 1998). Indeed for this case the equations (44) have forms:

$$\rho \frac{d}{dt}\dot{\vec{u}} - (\lambda + 2\mu)\nabla(\nabla \vec{u}) + \mu[\nabla[\nabla \vec{u}]] - \delta[\nabla \vec{\varphi}] = 0 \tag{44b}$$

$$I \frac{d}{dt}\dot{\vec{\varphi}} - \varepsilon \nabla(\nabla \vec{\varphi}) + \varsigma[\nabla[\nabla \vec{\varphi}]] + \sigma \vec{\varphi} + \delta[\nabla \vec{u}] = 0 \tag{45b}$$

The explicit form of these equations confirms that they are indeed the Cosserat continuum. If one sets formally $\delta = 0$, then equations (44b) and (45b) are split and the equation (44b) reduces to ordinal equation of the elasticity theory and the equation (45b) represents the wave equation for angular momentum.

When dissipation exists the system of equations (44)-(45) contains additional terms responsible for this dissipation

$$\rho \ddot{\vec{u}} - (\lambda + 2\mu)\nabla(\nabla \vec{u}) + \mu[\nabla[\nabla \vec{u}]] - \delta[\nabla \vec{\varphi}] = 0 \tag{44c}$$

$$I \ddot{\vec{\varphi}} - \varepsilon \nabla(\nabla \vec{\varphi}) + \varsigma[\nabla[\nabla \vec{\varphi}]] + \sigma \vec{\varphi} + \delta[\nabla \vec{u}] = -\gamma \dot{\vec{\varphi}} \tag{45c}$$

For the case $\varepsilon = 0$, $\varsigma = 0$ and $I = 0$ the second equation (45c) reduces to the pure relaxation form:

Generalized Variational Principle for Dissipative Hydrodynamics: Shear Viscosity from Angular Momentum Relaxation in the Hydrodynamical Description of Continuum Mechanics

63

$$\dot{\bar{\varphi}} = -\frac{\sigma}{\gamma}\bar{\varphi} - \frac{\delta}{\gamma}[\nabla\vec{u}] \tag{46}$$

Its solution can be represented in the form:

$$\bar{\varphi} = -\frac{\delta}{\gamma}\int_{-\infty}^{t} dt' e^{-\frac{\sigma}{\gamma}(t-t')}[\nabla\vec{u}] \tag{47a}$$

Substitution (47a) in (44c) leads to the following result

$$\rho\ddot{\vec{u}} - (\lambda + 2\mu)\nabla(\nabla\vec{u}) + \mu[\nabla[\nabla\vec{u}]] = -\frac{\delta^2}{\gamma}\int_{-\infty}^{t} dt' e^{-\frac{\sigma}{\gamma}(t-t')}[\nabla[\nabla\vec{u}]] \tag{48a}$$

For the case of large times $t\sigma/\gamma \gg 1$ the upper limit of integration gives the principal contribution and equation reduces to the form

$$\rho\ddot{\vec{u}} - (\lambda + 2\mu)\nabla(\nabla\vec{u}) + \left(\mu - \frac{\delta^2}{\sigma}\right)[\nabla[\nabla\vec{u}]] = \gamma\frac{\delta^2}{\sigma^2}[\nabla\dot{\vec{u}}] \tag{48b}$$

By the reason that the medium at large times should behave like a fluid then the following condition has to be satisfied

$$\mu - \frac{\delta^2}{\sigma} = 0 \tag{49}$$

Taking into account condition (49) let's make more accurate estimation of the integral, computing it by parts

$$\rho\ddot{\vec{u}} - (\lambda + 2\mu)\nabla(\nabla\vec{u}) = -\frac{\delta^2}{\sigma}\int_{-\infty}^{t} dt' e^{-\frac{\sigma}{\gamma}(t-t')}[\nabla[\nabla\dot{\vec{u}}]] \tag{48c}$$

The corresponding estimation for the large time limit $t \gg \gamma/\sigma$ reduces to the equation

$$\rho\ddot{\vec{u}} - (\lambda + 2\mu)\nabla(\nabla\vec{u}) = \gamma\frac{\mu^2}{\delta^2}[\nabla[\nabla\dot{\vec{u}}]] \tag{48d}$$

which coincides with the structure of Navier-Stokes equation in the presence of shear viscosity.

Let's consider the case with non zero moment of inertia $I \neq 0$. For this case the second equation (45c) is also local in space and it can be resolved for the function $\bar{\varphi}$ in Fourier representation ($t \rightarrow \omega$)

$$\bar{\varphi} = \frac{-\delta}{-I\omega^2 + i\omega\gamma + \sigma}[\nabla\vec{u}] \tag{50}$$

The zeros of the denominator

$$i\omega_{1,2} = \frac{1}{2I}\left(-\gamma \pm \sqrt{\gamma^2 - 4\sigma I}\right) \tag{51}$$

determine two modes of angular momentum relaxation. Under condition $I < \gamma^2/(4\sigma)$ both zeros are real and have the following asymptotics for small momentum of inertia $I \to 0$:

$$i\omega_1 \approx -\frac{\sigma}{\gamma} \qquad i\omega_2 \approx -\frac{\gamma}{I} \qquad\qquad (52)$$

The first zero does not depend on momentum of inertia I and the second root goes to infinity when $I \to 0$. Under condition $I = \gamma^2/(4\sigma)$ the zeros coincide and have the value $i\omega_1 \approx -2\frac{\sigma}{\gamma}$, and under the condition $I > \gamma^2/(4\sigma)$ the zeros are complex conjugated with negative real part, which decreases with increase of I. The last case corresponds to the resonant relaxation of angular momentum.

In the time representation the solution of the equation (50) can be written in the form

$$\vec{\varphi} = -\int_{-\infty}^{t} dt' e^{-\frac{\gamma}{2I}(t-t')} [\nabla \vec{u}] \left\{ \frac{2\delta}{\sqrt{\ldots}} sh\left(\frac{\sqrt{\ldots}}{2I}(t - t') \right) \right\} \qquad\qquad (47b)$$

here the notation $\sqrt{\ldots} = \sqrt{\gamma^2 - 4\sigma I}$ is used. For the case of resonant relaxation $I > \gamma^2/(4\sigma)$ the corresponding expression has the form

$$\vec{\varphi} = -\int_{-\infty}^{t} dt' e^{-\frac{\gamma}{2I}(t-t')} [\nabla \vec{u}] \left\{ \frac{2\delta}{\sqrt{|\ldots|}} \sin\left(\frac{\sqrt{|\ldots|}}{2I}(t - t') \right) \right\} \qquad\qquad (47c)$$

Substitution of the explicit expressions (47b) or (47c) in the equation (44c) gives the generalisation of the Navier – Stokes equation for a solid medium with local relaxation of angular momentum. As was mentioned above under special condition (49) and in the limiting case (52) this equation reduces exactly to the form of Navier – Stokes equation.

Thus, it is shown that relaxation of angular momentum of material points consisting a continuum can be considered as physical reason for appearance of terms with shear viscosity in Navier-Stokes equation. Without dissipation additional degree of freedom dealt with angular momentum leads to the well known Cosserat continuum.

4. Conclusion

The first part of the chapter presents an original formulation of the generalized variational principle (GVP) for dissipative hydrodynamics (continuum mechanics) as a direct combination of Hamilton's and Onsager's variational principles. The GVP for dissipative continuum mechanics is formulated as Hamilton's variational principle in terms of two independent field variables i.e. the mean mass and the heat displacement fields. It is important to mention that existence of two independent fields gives us opportunity to consider a closed mechanical system and hence to formulate variational principle. Dissipation plays only a role of energy transfer between the mean mass and the heat displacement fields. A system of equations for these fields is derived from the extreme condition for action with a Lagrangian density in the form of the difference between the kinetic and the free energies minus the time integral of the dissipation function. All mentioned potential functions are considered as a general positively determined quadratic forms of time or space derivatives of the mean mass and the heat displacement fields. The generalized system of hydrodynamical equations is then evaluated on the basis of the GVP. At low frequencies this system corresponds to the traditional Navier – Stokes equation system. It allowed us to determine all coefficients of quadratic forms by direct comparison with the Navier – Stokes equation system.

The second part of the chapter is devoted to consistent introduction of viscous terms into the equation of fluid motion on the basis of the GVP. A tensor internal parameter is used for description of relaxation processes in vicinity of quasi-equilibrium state by analogy with the Mandelshtam – Leontovich approach. The derived equation of motion describes the viscosity relaxation phenomenon and generalizes the well known Navier – Stokes equation. At low frequencies the equation of fluid motion reduces exactly to the form of Navier – Stokes equation. Nevertheless there is still a question about physical interpretation of the used internal parameter. The answer on this question is presented in the last section of the chapter.

It is shown that the internal parameter responsible for shear viscosity can be interpreted as a consequence of relaxation of angular momentum of material points constituting a mechanical continuum. Due to angular momentum balance law the rotational degree of freedom as independent variable appears additionally to the mean mass displacement field. For the dissipationless case this approach leads to the well-known Cosserat continuum. When dissipation prevails over momentum of inertion this approach describes local relaxation of angular momentum and corresponds to the sense of the internal parameter. It is important that such principal parameter of Cosserat continuum as the inertia moment of intrinsic microstructure can completely vanish from the description for dissipative continuum. The independent equation of motion for angular momentum in this case reduces to local relaxation and after its substitution into the momentum balance equation leads to the viscous terms in Navier – Stokes equation. Thus, it is shown that the nature of viscosity phenomenon can be interpreted as relaxation of angular momentum of material points on the kinetic level.

5. Acknowledgment

The work was supported by ISTC grant 3691 and by RFBR grant №09-02-00927-a.

6. References

Derdichevsky V.L. (2009). *Variational principles of continuum mechanics*, Springer-Verlag, ISBN 978-3-540-88466-8, Berlin.

Biot M. (1970). *Variational principles in heat transfer*. Oxford, University Press.

Deresiewicz H. (1957). Plane wave in a thermoplastic solids. *The Journal of the Acoustcal Society of America*, Vol.29, pp.204-209, ISSN 0001-4966.

Erofeev V.I. (1998). *Wave processes in solids with microstructure*, Moscow State University, Moscow.

Glensdorf P., Prigogine I., (1971). *Thermodynamic Theory of Structure, Stability, and Fluctuations*, Wiley, New York.

Gyarmati I. (1970). *Non-equilibrium thermodynamics. Field theory and variational principles*. Berlin, Springer-Verlag.

Kunin I.A. (1975). *Theory of elastic media with micro structure* , Nauka, Moscow.

Landau L.D., Lifshitz E.M. (1986). *Theoretical physics. Vol.6. Hydrodynamics*, Nauka, Moscow.

Landau L.D., Lifshitz E.M. (1972). *Theoretical physics. Vol.7. Theory of elasticity*, Nauka, Moscow.

Landau L.D., Lifshitz E.M. (1964). *Theoretical physics. Vol.5. Statistical physics.* Nauka, Moscow.

Lykov A.V. (1967). *Theory of heat conduction,* Moscow, Vysshaya Shkola.

Mandelshtam L.I., Leontovich M.A. (1937). To the sound absorption theory in liquids, *The Journal of Experimental and Theoretical Physics*, Vol.7, No.3, pp. 438-444, ISSN 0044-4510 (in Russian).

Martynov G.A. (2001). Hydrodynamic theory of sound wave propagation. *Theoretical and Mathematical Physics*, Vol.129, pp.1428-1438, ISSN 0564-6162.

Maximov G.A. (2006). On the variational principle for dissipative hydrodynamics. *Preprint 006-2006, Moscow Engineering Physics Institute*, Moscow. (in Russian)

Maximov G.A. (2008). Generalized variational principle for dissipative hydrodynamics and its application to the Biot's equations for multicomponent, multiphase media with temperature gradient, In: *New Research in Acoustics*, B.N. Weis, (Ed.), 21-61, Nova Science Publishers Inc., ISBN 978-1-60456-403-7.

Maximov G.A. (2010). Generalized variational principle for dissipative hydrodynamics and its application to the Biot's theory for the description of a fluid shear relaxation, *Acta Acustica united with Acustica*, Vol.96, pp. 199-207, ISSN 1610-1928.

Nettleton R.E. (1960). Relaxation theory of thermal conduction in liquids. *Physics of Fluids*, Vol.3, pp.216-223, ISSN 1070-6631

Novatsky V. (1975). *Theory of elasticity*, Mir, Moscow.

Onsager L. (1931a). Reciprocal relations in irreversible process I. *Physical Review*, Vol.37, pp.405-426.

Onsager L. (1931b). Reciprocal relations in irreversible process II. *Physical Review*, Vol. 38, p.2265-2279.

Shliomis M.I. (1966). Hydrodynamics of a fluid with intrinsic rotation, *The Journal of Experimental and Theoretical Physics*, Vol.51, No.7, pp.258-265, ISSN 0044-4510 (in Russian).

Sorokin V.S. (1943). On internal friction of liquids and gases possessed hidden angular momentum, *The Journal of Experimental and Theoretical Physics*, Vol.13, No.7-8, pp. 306-312, ISSN 0044-4510 (in Russian).

Zhdanov V.M., Roldugin V.I. (1998). Non-equilibrium thermodynamics and kinetic theory of rarefied gases. *Physics-Uspekh,*. Vol.41, No.4, pp. 349-381, ISSN 0042-1294.

4

Nonautonomous Solitons: Applications from Nonlinear Optics to BEC and Hydrodynamics

T. L. Belyaeva[1] and V. N. Serkin[2]

[1]*Universidad Autónoma del Estado de México*
[2]*Benemerita Universidad Autónoma de Puebla*
Mexico

1. Introduction

Nonlinear science is believed by many outstanding scientists to be the most deeply important frontier for understanding Nature (Christiansen et al., 2000; Krumhansl, 1991). The interpenetration of main ideas and methods being used in different fields of science and technology has become today one of the decisive factors in the progress of science as a whole. Among the most spectacular examples of such an interchange of ideas and theoretical methods for analysis of various physical phenomena is the problem of solitary wave formation in nonautonomous and inhomogeneous dispersive and nonlinear systems. These models are used in a variety of fields of modern nonlinear science from hydrodynamics and plasma physics to nonlinear optics and matter waves in Bose-Einstein condensates.

The purpose of this Chapter is to show the progress that is being made in the field of the exactly integrable nonautonomous and inhomogeneous nonlinear evolution equations possessing the exact soliton solutions. These kinds of solitons in nonlinear nonautonomous systems are well known today as nonautonomous solitons. Most of the problems considered in the present Chapter are motivated by their practical significance, especially the hydrodynamics applications and studies of possible scenarios of generations and controlling of monster (rogue) waves by the action of different nonautonomous and inhomogeneous external conditions.

Zabusky and Kruskal (Zabusky & Kruskal, 1965) introduced for the first time the soliton concept to characterize nonlinear solitary waves that do not disperse and preserve their identity during propagation and after a collision. The Greek ending "on" is generally used to describe elementary particles and this word was introduced to emphasize the most remarkable feature of these solitary waves. This means that the energy can propagate in the localized form and that the solitary waves emerge from the interaction completely preserved in form and speed with only a phase shift. Because of these defining features, the classical soliton is being considered as the ideal natural data bit. It should be emphasized that today, the optical soliton in fibers presents a beautiful example in which an abstract mathematical concept has produced a large impact on the real world of high technologies (Agrawal, 2001; Akhmediev, 1997; 2008; Dianov et al., 1989; Hasegawa, 1995; 2003; Taylor, 1992).

Solitons arise in any physical system possessing both nonlinearity and dispersion, diffraction or diffusion (in time or/and space). The classical soliton concept was developed for nonlinear and dispersive systems that have been autonomous; namely, time has only played the role of the independent variable and has not appeared explicitly in the nonlinear evolution equation. A not uncommon situation is one in which a system is subjected to some form of external time-dependent force. Such situations could include repeated stress testing of a soliton in nonuniform media with time-dependent density gradients.

Historically, the study of soliton propagation through density gradients began with the pioneering work of Tappert and Zabusky (Tappert & Zabusky, 1971). As early as in 1976 Chen and Liu (Chen, 1976; 1978) substantially extended the concept of classical solitons to the accelerated motion of a soliton in a linearly inhomogeneous plasma. It was discovered that for the nonlinear Schrödinger equation model (NLSE) with a linear external potential, the inverse scattering transform (IST) method can be generalized by allowing the time-varying eigenvalue (TVE), and as a consequence of this, the solitons with time-varying velocities (but with time invariant amplitudes) have been predicted (Chen, 1976; 1978). At the same time Calogero and Degaspieris (Calogero, 1976; 1982) introduced a general class of soliton solutions for the nonautonomous Korteweg-de Vries (KdV) models with varying nonlinearity and dispersion. It was shown that the basic property of solitons, to interact elastically, was also preserved, but the novel phenomenon was demonstrated, namely the fact that each soliton generally moves with variable speed as a particle acted by an external force rather than as a free particle (Calogero, 1976; 1982). In particular, to appreciate the significance of this analogy, Calogero and Degaspieris introduced the terms boomeron and trappon instead of classical KdV solitons (Calogero, 1976; 1982). Some analytical approaches for the soliton solutions of the NLSE in the nonuniform medium were developed by Gupta and Ray (Gupta, 1981), Herrera (Herrera, 1984), and Balakrishnan (Balakrishnan, 1985). More recently, different aspects of soliton dynamics described by the nonautonomous NLSE models were investigated in (Serkin & Hasegawa, 2000a;b; 2002; Serkin et al., 2004; 2007; 2001a;b). In these works, the "ideal" soliton-like interaction scenarios among solitons have been studied within the generalized nonautonomous NLSE models with varying dispersion, nonlinearity and dissipation or gain. One important step was performed recently by Serkin, Hasegawa and Belyaeva in the Lax pair construction for the nonautonomous nonlinear Schrödinger equation models (Serkin et al., 2007). Exact soliton solutions for the nonautonomous NLSE models with linear and harmonic oscillator potentials substantially extend the concept of classical solitons and generalize it to the plethora of nonautonomous solitons that interact elastically and generally move with varying amplitudes, speeds and spectra adapted both to the external potentials and to the dispersion and nonlinearity variations. In particular, solitons in nonautonomous physical systems exist only under certain conditions and varying in time nonlinearity and dispersion cannot be chosen independently; they satisfy the exact integrability conditions. The law of soliton adaptation to an external potential has come as a surprise and this law is being today the object of much concentrated attention in the field. The interested reader can find many important results and citations, for example, in the papers published recently by Zhao et al. (He et al., 2009; Luo et al., 2009; Zhao et al., 2009; 2008), Shin (Shin, 2008) and (Kharif et al., 2009; Porsezian et al., 2007; Yan, 2010).

How can we determine whether a given nonlinear evolution equation is integrable or not? The ingenious method to answer this question was discovered by Gardner, Green, Kruskal and Miura (GGKM) (Gardner et al., 1967). Following this work, Lax (Lax, 1968) formulated a general principle for associating of nonlinear evolution equations with linear operators, so that the eigenvalues of the linear operator are integrals of the nonlinear equation. Lax developed the method of inverse scattering transform (IST) based on an abstract formulation of evolution equations and certain properties of operators in a Hilbert space, some of which are well known in the context of quantum mechanics. Ablowitz, Kaup, Newell, Segur (AKNS) (Ablowitz et al., 1973) have found that many physically meaningful nonlinear models can be solved by the IST method.

In the traditional scheme of the IST method, the spectral parameter Λ of the auxiliary linear problem is assumed to be a time independent constant $\Lambda'_t = 0$, and this fact plays a fundamental role in the development of analytical theory (Zakharov, 1980). The nonlinear evolution equations that arise in the approach of variable spectral parameter, $\Lambda'_t \neq 0$, contain, as a rule, some coefficients explicitly dependent on time. The IST method with variable spectral parameter makes it possible to construct not only the well-known models for nonlinear autonomous physical systems, but also discover many novel integrable and physically significant nonlinear nonautonomous equations.

In this work, we clarify our algorithm based on the Lax pair generalization and reveal generic properties of nonautonomous solitons. We consider the generalized nonautonomous NLSE and KdV models with varying dispersion and nonlinearity from the point of view of their exact integrability. It should be stressed that to test the validity of our predictions, the experimental arrangement should be inspected to be as close as possible to the optimal map of parameters, at which the problem proves to be exactly integrable (Serkin & Hasegawa, 2000a;b; 2002). Notice, that when Serkin and Hasegawa formulated their concept of solitons in nonautonomous systems (Serkin & Hasegawa, 2000a;b; 2002), known today as nonautonomous solitons and SH-theorems (Serkin & Hasegawa, 2000a;b; 2002) published for the first time in (Serkin & Hasegawa, 2000a;b; 2002), they emphasized that "the methodology developed provides for a systematic way to find an infinite number of the novel stable bright and dark "soliton islands" in a "sea of solitary waves" with varying dispersion, nonlinearity, and gain or absorption" (Belyaeva et al., 2011; Serkin et al., 2010a;b). The concept of nonautonomous solitons, the generalized Lax pair and generalized AKNS methods described in details in this Chapter can be applied to different physical systems, from hydrodynamics and plasma physics to nonlinear optics and matter-waves and offer many opportunities for further scientific studies. As an illustrative example, we show that important mathematical analogies between different physical systems open the possibility to study optical rogue waves and ocean rogue waves in parallel and, due to the evident complexity of experiments with rogue waves in open oceans, this method offers remarkable possibilities in studies nonlinear hydrodynamic problems by performing experiments in the nonlinear optical systems with nonautonomous solitons and optical rogue waves.

2. Lax operator method and exact integrability of nonautonomous nonlinear and dispersive models with external potentials

The classification of dynamic systems into autonomous and nonautonomous is commonly used in science to characterize different physical situations in which, respectively, external time-dependent driving force is being present or absent. The mathematical treatment of nonautonomous system of equations is much more complicated then of traditional autonomous ones. As a typical illustration we may mention both a simple pendulum whose length changes with time and parametrically driven nonlinear Duffing oscillator (Nayfeh & Balachandran, 2004).

In the framework of the IST method, the nonlinear integrable equation arises as the compatibility condition of the system of the linear matrix differential equations

$$\psi_x = \widehat{\mathcal{F}}\psi(x,t), \qquad \psi_t = \widehat{\mathcal{G}}\psi(x,t). \tag{1}$$

Here $\psi(x,t) = \{\psi_1, \psi_2\}^T$ is a 2-component complex function, $\widehat{\mathcal{F}}$ and $\widehat{\mathcal{G}}$ are complex-valued (2×2) matrices. Let us consider the general case of the IST method with a time-dependent spectral parameter $\Lambda(T)$ and the matrices $\widehat{\mathcal{F}}$ and $\widehat{\mathcal{G}}$

$$\widehat{\mathcal{F}}(\Lambda; S, T) = \widehat{\mathcal{F}} \left\{ \Lambda(T), q\,[S(x,t), T]; \frac{\partial q}{\partial S}\left(\frac{\partial S}{\partial x}\right); \frac{\partial^2 q}{\partial S^2}\left(\frac{\partial S}{\partial x}\right)^2; ...; \frac{\partial^n q}{\partial S^n}\left(\frac{\partial S}{\partial x}\right)^n \right\}$$

$$\widehat{\mathcal{G}}(\Lambda; S, T) = \widehat{\mathcal{G}} \left\{ \Lambda(T), q\,[S(x,t), T]; \frac{\partial q}{\partial S}\left(\frac{\partial S}{\partial x}\right); \frac{\partial^2 q}{\partial S^2}\left(\frac{\partial S}{\partial x}\right)^2; ...; \frac{\partial^n q}{\partial S^n}\left(\frac{\partial S}{\partial x}\right)^n \right\},$$

dependent on the generalized coordinates $S = S(x,t)$ and $T(t) = t$, where the function $q\,[S(x,t), T]$ and its derivatives denote the scattering potentials $Q(S,T)$ and $R(S,T)$ and their derivatives, correspondingly. The condition for the compatibility of the pair of linear differential equations (1) takes a form

$$\frac{\partial \widehat{\mathcal{F}}}{\partial T} + \frac{\partial \widehat{\mathcal{F}}}{\partial S}S_t - \frac{\partial \widehat{\mathcal{G}}}{\partial S}S_x + \left[\widehat{\mathcal{F}}, \widehat{\mathcal{G}}\right] = 0, \tag{2}$$

where

$$\widehat{\mathcal{F}} = -i\Lambda(T)\widehat{\sigma}_3 + \widehat{\mathcal{U}}\widehat{\phi}, \tag{3}$$

$$\widehat{\mathcal{G}} = \begin{pmatrix} A & B \\ C & -A \end{pmatrix}, \tag{4}$$

$\widehat{\sigma}_3$ is the Pauli spin matrix and matrices $\widehat{\mathcal{U}}$ and $\widehat{\phi}$ are given by

$$\widehat{\mathcal{U}} = \sqrt{\sigma}F^{\gamma}(T)\begin{pmatrix} 0 & Q(S,T) \\ R(S,T) & 0 \end{pmatrix}, \tag{5}$$

$$\widehat{\phi} = \begin{pmatrix} \exp[-i\varphi/2] & 0 \\ 0 & \exp[i\varphi/2] \end{pmatrix}. \tag{6}$$

Here $F(T)$ and $\varphi(S,T)$ are real unknown functions, γ is an arbitrary constant, and $\sigma = \pm 1$. The desired elements of $\widehat{\mathcal{G}}$ matrix (known in the modern literature as the AKNS elements) can be constructed in the form $\widehat{\mathcal{G}} = \sum_{k=0}^{k=3} G_k \Lambda^k$, with time varying spectral parameter given by

$$\Lambda_T = \lambda_0(T) + \lambda_1(T)\Lambda(T), \tag{7}$$

where time-dependent functions $\lambda_0(T)$ and $\lambda_1(T)$ are the expansion coefficients of Λ_T in powers of the spectral parameter $\Lambda(T)$.
Solving the system (2-6), we find both the matrix elements A, B, C

$$A = -i\lambda_0 S/S_x + a_0 - \frac{1}{4}a_3\sigma F^{2\gamma}(QR\varphi_S S_x + iQR_S S_x - iRQ_S S_x) \tag{8}$$

$$+ \frac{1}{2}a_2\sigma F^{2\gamma}QR + \Lambda\left(-i\lambda_1 S/S_x + \frac{1}{2}a_3\sigma F^{2\gamma}QR + a_1\right) + a_2\Lambda^2 + a_3\Lambda^3,$$

$$B = \sqrt{\sigma}F^{\gamma}\exp[i\varphi S/2]\{-\frac{i}{4}a_3 S_x^2\left(Q_{SS} + \frac{i}{2}Q\varphi_{SS} - \frac{1}{4}Q\varphi_S^2 + iQ_S\varphi_S\right)$$

$$- \frac{i}{4}a_2 Q\varphi_S S_x - \frac{1}{2}a_2 Q_S S_x + iQ\left(-i\lambda_1 S/S_x + \frac{1}{2}a_3\sigma F^{2\gamma}QR + a_1\right)$$

$$+ \Lambda\left(-\frac{i}{4}a_3 Q\varphi_S S_x - \frac{1}{2}a_3 Q_S S_x + ia_2 Q\right) + ia_3\Lambda^2 Q\},$$

$$C = \sqrt{\sigma} F^{\gamma} \exp[-i\varphi S/2] \{ -\frac{i}{4} a_3 S_x^2 \left(R_{SS} - \frac{i}{2} R\varphi_{SS} - \frac{1}{4} R\varphi_S^2 - iR_S\varphi_S \right)$$

$$-\frac{i}{4} a_2 R\varphi_S S_x + \frac{1}{2} a_2 R_S S_x + iR \left(-i\lambda_1 S/S_x + \frac{1}{2} a_3 \sigma F^{2\gamma} QR + a_1 \right)$$

$$+\Lambda \left(-\frac{i}{4} a_3 R\varphi_S S_x + \frac{1}{2} a_3 R_S S_x + ia_2 R \right) + ia_3 \Lambda^2 R \},$$

and two general equations

$$iQ_T = \frac{1}{4} a_3 Q_{SSS} S_x^3 + \frac{3i}{8} a_3 Q_{SS} \varphi_S S_x^3 - \frac{3i}{4} a_3 \sigma F^{2\gamma} Q^2 R\varphi_S S_x \tag{9}$$

$$-\frac{3}{2} a_3 \sigma F^{2\gamma} QR Q_S S_x - \frac{i}{2} a_2 Q_{SS} S_x^2 + ia_2 \sigma F^{2\gamma} Q^2 R$$

$$+iQ_S \left(-S_t + \lambda_1 S + ia_1 S_x - \frac{i}{2} a_2 \varphi_S S_x^2 + \frac{3}{8} a_3 \varphi_{SS} S_x^3 + \frac{3i}{16} a_3 \varphi_S^2 S_x^3 \right)$$

$$+Q \left(i\lambda_1 - i\gamma \frac{F_T}{F} + \frac{1}{2} a_2 \varphi_{SS} S_x^2 - \frac{3}{16} a_3 \varphi_S \varphi_{SS} S_x^3 \right)$$

$$+Q \left[2\lambda_0 S/S_x + 2ia_0 + \frac{1}{2} (\varphi_T + \varphi_S S_t) - \frac{1}{2} \lambda_1 S\varphi_S - \frac{i}{2} a_1 \varphi_S S_x \right]$$

$$+Q \left(\frac{i}{8} a_2 \varphi_S^2 S_x^2 - \frac{i}{32} a_3 \varphi_S^3 S_x^3 + \frac{i}{8} a_3 \varphi_{SSS} S_x^3 \right)$$

$$iR_T = \frac{1}{4} a_3 R_{SSS} S_x^3 - \frac{3i}{8} a_3 R_{SS} \varphi_S S_x^3 + \frac{3i}{4} a_3 \sigma F^{2\gamma} R^2 Q\varphi_S S_x \tag{10}$$

$$-\frac{3}{2} a_3 \sigma F^{2\gamma} R^2 Q_S S_x + \frac{i}{2} a_2 R_{SS} S_x^2 - ia_2 \sigma F^{2\gamma} R^2 Q$$

$$+iR_S \left(-S_t + \lambda_1 S + ia_1 S_x - \frac{i}{2} a_2 \varphi_S S_x^2 - \frac{3}{8} a_3 \varphi_{SS} S_x^3 + \frac{3i}{16} a_3 \varphi_S^2 S_x^3 \right)$$

$$+R \left(i\lambda_1 - i\gamma \frac{F_T}{F} + \frac{1}{2} a_2 \varphi_{SS} S_x^2 - \frac{3}{16} a_3 \varphi_S \varphi_{SS} S_x^3 \right)$$

$$+R \left[-2\lambda_0 S/S_x - 2ia_0 - \frac{1}{2} (\varphi_T + \varphi_S S_t) + \frac{1}{2} \lambda_1 S\varphi_S + \frac{i}{2} a_1 \varphi_S S_x \right]$$

$$+R \left(-\frac{i}{8} a_2 \varphi_S^2 S_x^2 + \frac{i}{32} a_3 \varphi_S^3 S_x^3 - \frac{i}{8} a_3 \varphi_{SSS} S_x^3 \right),$$

where the arbitrary time-dependent functions $a_0(T)$, $a_1(T)$, $a_2(T)$, $a_3(T)$ have been introduced within corresponding integrations.

By using the following reduction procedure $R = -Q^*$, it is easy to find that two equations (9) and (10) take the same form if the following conditions

$$a_0 = -a_0^*, \ a_1 = -a_1^*, \ a_2 = -a_2^*, \ a_3 = -a_3^*, \tag{11}$$

$$\lambda_0 = \lambda_0^*, \quad \lambda_1 = \lambda_1^*, \quad F = F^*$$

are fulfilled.

3. Generalized nonlinear Schrödinger equation and solitary waves in nonautonomous nonlinear and dispersive systems: nonautonomous solitons

Let us study a special case of the reduction procedure for Eqs. (9,10) when $a_3 = 0$

$$A = -i\lambda_0 S/S_x + a_0(T) - \frac{1}{2}a_2(T)\sigma F^{2\gamma}|Q|^2 - i\lambda_1 S/S_x\Lambda + a_1(T)\Lambda + a_2(T)\Lambda^2,$$

$$B = \sqrt{\sigma}F^\gamma \exp(i\varphi/2)\left\{-\frac{i}{4}a_2(T)Q\varphi_S S_x - \frac{1}{2}a_2(T)Q_S S_x\right\} +$$
$$i\{Q[-i\lambda_1 S/S_x + a_1(T) + \Lambda a_2(T)]\},$$

$$C = \sqrt{\sigma}F^\gamma \exp(-i\varphi/2)\left\{\frac{i}{4}a_2(T)Q^*\varphi_S S_x - \frac{1}{2}a_2(T)Q_S^* S_x\right\}$$
$$-i\{Q^*[-i\lambda_1 x + a_1(T) + \Lambda a_2(T)]\}.$$

In accordance with conditions (11), the imaginary functions $a_0(T)$, $a_1(T)$, $a_2(T)$ can be defined in the following way: $a_0(T) = i\gamma_0(T)$, $a_1(T) = iV(T)$, $a_2(T) = -iD_2(T)$, $R_2(T) = F^{2\gamma}D_2(T)$, where $D_2(T)$, $V(T)$, $\gamma_0(T)$ are arbitrary real functions. The coefficients $D_2(T)$ and $R_2(T)$ are represented by positively defined functions (for $\sigma = -1$, γ is assumed as a semi-entire number).

Then, Eqs. (9,10) can be transformed into

$$iQ_T = -\frac{1}{2}D_2 Q_{SS}S_x^2 - \sigma R_2|Q|^2 Q - i\widetilde{V}Q_S + i\Gamma Q + UQ, \tag{12}$$

where

$$\widetilde{V}(S,T) = \frac{1}{2}D_2 S_x^2\varphi_S + VS_x + S_t - \lambda_1 S,$$

$$U(S,T) = \frac{1}{8}D_2 S_x^2\varphi_S^2 - 2\gamma_0 + \frac{1}{2}(\varphi_T + \varphi_S S_t + VS_x\varphi_S) + 2\lambda_0 S/S_x - \frac{1}{2}\lambda_1\varphi_S S, \tag{13}$$

$$\Gamma = \left(-\gamma\frac{F_T}{F} - \frac{1}{4}D_2 S_x^2\varphi_{SS} + \lambda_1\right) = \left(\frac{1}{2}\frac{W(R_2, D_2)}{R_2 D_2} - \frac{1}{4}D_2 S_x^2\varphi_{SS} + \lambda_1\right). \tag{14}$$

Eq.(12) can be written down in the independent variables (x,t)

$$iQ_t + \frac{1}{2}D_2(t)Q_{xx} + \sigma R_2(t)|Q|^2 Q - U(x,t)Q + i\widetilde{V'}Q_x = i\Gamma(t)Q. \tag{15}$$

Let us transform Eq.(15) into the more convenient form

$$iQ_t + \frac{1}{2}D_2 Q_{xx} + \sigma R_2|Q|^2 Q - UQ = i\Gamma Q \tag{16}$$

using the following condition

$$\widetilde{V'} = \frac{1}{2}D_2 S_x\varphi_S + V - \lambda_1 S/S_x = 0. \tag{17}$$

If we apply the commonly accepted in the IST method (Ablowitz et al., 1973) reduction: $V = -ia_1 = 0$, we find a parameter λ_1 from (17)

$$\lambda_1 = \frac{1}{2}D_2 S_x^2\varphi_S/S, \tag{18}$$

and the corresponding potential $U(S,T)$ from Eq.(13):

$$U(S,T) = -2\gamma_0 + 2\lambda_0 S/S_x + \frac{1}{2}(\varphi_T + \varphi_S S_t) - \frac{1}{8}D_2 S_x^2 \varphi_S^2. \tag{19}$$

According to Eq.(14), the gain or absorption coefficient now is represented by

$$\Gamma = \frac{1}{2}\frac{W(R_2, D_2)}{R_2 D_2} - \frac{1}{4}D_2 S_x^2 \varphi_{SS} + \frac{1}{2}D_2 S_x^2 \varphi_S / S. \tag{20}$$

Let us consider some special choices of variables to specify the solutions of (16). First of all, we assume that variables are factorized in the phase profile $\varphi(S,T)$ as $\varphi = C(T)S^\alpha$. The first term in the real potential (19) represents some additional time-dependent phase $e^{2\gamma_0(t)t}$ of the solution $Q(x,t)$ for the equation (16) and, without loss of the generality, we use $\gamma_0 = 0$. The second term in (19) depends linearly on S. The NLSE with the linear spatial potential and constant λ_0, describing the case of Alfen waves propagation in plasmas, has been studied previously in Ref. (Chen, 1976). We will study the more general case of chirped solitons in the Section 4 of this Chapter. Now, taking into account three last terms in (19), we obtain

$$U(S,T) = 2\lambda_0 S/S_x + \frac{1}{2}C_T S^\alpha + 1/2\alpha C S^{\alpha-1} S_t - \frac{1}{8}D_2 C^2 S_x^2 \alpha^2 S^{2\alpha-2}. \tag{21}$$

The gain or absorption coefficient (20) becomes

$$\Gamma(T) = \frac{1}{2}\frac{W(R_2, D_2)}{R_2 D_2} + \frac{\alpha}{4}(3-\alpha)D_2 S_x^2 C S^{\alpha-2} \tag{22}$$

and Eq.(18) takes a form

$$\lambda_1 = \frac{1}{2}D_2 S_x^2 C\alpha S^{\alpha-2}. \tag{23}$$

If we assume that the functions $\Gamma(T)$ and $\lambda_1(T)$ depend only on T and do not depend on S, we conclude that $\alpha = 0$ or $\alpha = 2$.

The study of the soliton solutions of the nonautonomous NLSE with varying coefficients without time and space phase modulation (chirp) and corresponding to the case of $\alpha = 0$ has been carried out in Ref. (Serkin & Belyaeva, 2001a;b). Let us find here the solutions of Eq.(16) with chirp in the case of $\alpha = 2$, $\varphi(S,T) = C(T)S^2$. In this case, Eq. (18) becomes $\lambda_1 = D_2 S_x^2 C$. Now, the real spatial-temporal potential (21) takes the form

$$U[S(x,t),T)] = 2\lambda_0 S/S_x + \frac{1}{2}\left(C_T - D_2 S_x^2 C^2\right)S^2 + CSS_t$$

Consider the simplest option to choose the variable $S(x,t)$ when the variables (x,t) are factorized: $S(x,t) = P(t)x$. In this case, all main characteristic functions: the phase modulation

$$\varphi(x,t) = \Theta(t)x^2, \tag{24}$$

the real potential

$$U(x,t) = 2\lambda_0 x + \frac{1}{2}\left(\Theta_t - D_2 \Theta^2\right)x^2 \equiv 2\lambda_0(t)x + \frac{1}{2}\Omega^2(t)x^2, \tag{25}$$

the gain (or absorption) coefficient

$$\Gamma(t) = \frac{1}{2}\left(\frac{W(R_2, D_2)}{R_2 D_2} + D_2 P^2 C\right) = \frac{1}{2}\left(\frac{W(R_2, D_2)}{R_2 D_2} + D_2 \Theta\right) \tag{26}$$

and the spectral parameter λ_1

$$\lambda_1(t) = D_2 P^2 C = D_2(t)\Theta(t) \tag{27}$$

are found to be dependent on the self-induced soliton phase shift $\Theta(t)$. Notice that the definition $\Omega^2(t) \equiv \Theta_t - D_2\Theta^2$ has been introduced in Eq.(25).

Now we can rewrite the generalized NLSE (16) with time-dependent nonlinearity, dispersion and gain or absorption in the form of the nonautonomous NLSE with linear and parabolic potentials

$$iQ_t + \frac{1}{2}D_2(t)Q_{xx} + \sigma R_2(t)\,|Q|^2\,Q - 2\lambda_0(t)x - \frac{1}{2}\,\Omega^2(t)x^2 Q = i\Gamma Q. \tag{28}$$

4. Hidden features of the soliton adaptation law to external potentials: the generalized Serkin-Hasegawa theorems

It is now generally accepted that solitary waves in nonautonomous nonlinear and dispersive systems can propagate in the form of so-called nonautonomous solitons or solitonlike similaritons (see (Atre et al., 2006; Avelar et al., 2009; Belić et al., 2008; Chen et al., 2007; Hao, 2008; He et al., 2009; Hernandez et al., 2005; Hernandez-Tenorio et al., 2007; Liu et al., 2008; Porsezian et al., 2009; 2007; Serkin et al., 2007; Shin, 2008; Tenorio et al., 2005; Wang et al., 2008; Wu, Li & Zhang, 2008; Wu, Zhang, Li, Finot & Porsezian, 2008; Zhang et al., 2008; Zhao et al., 2009; 2008) and references therein). Nonautonomous solitons interact elastically and generally move with varying amplitudes, speeds and spectra adapted both to the external potentials and to the dispersion and nonlinearity variations. The existence of specific laws of soliton adaptation to external gain and loss potentials was predicted by Serkin and Hasegawa in 2000 (Serkin & Hasegawa, 2000a;b; 2002). The physical mechanism resulting in the soliton stabilization in nonautonomous and dispersive systems was revealed in this paper. From the physical point of view, the adaptation means that solitons remain self similar and do not emit dispersive waves both during their interactions with external potentials and with each other. The soliton adaptation laws are known today as the Serkin-Hasegawa theorems (SH theorems). Serkin and Hasegawa obtained their SH-theorems by using the symmetry reduction methods when the initial nonautonomous NLSE can be transformed by the canonical autonomous NLSE under specific conditions found in (Serkin & Hasegawa, 2000a;b). Later, SH-theorems have been confirmed by different methods, in particular, by the Painleve analysis and similarity transformations (Serkin & Hasegawa, 2000a;b; 2002; Serkin et al., 2004; 2007; 2001a;b).

Substituting the phase profile $\Theta(t)$ given by Eq. (26) into Eq. (25), it is straightforward to verify that the frequency of the harmonic potential $\Omega(t)$ is related with dispersion $D_2(t)$, nonlinearity $R_2(t)$ and gain or absorption coefficient $\Gamma(t)$ by the following conditions

$$\Omega^2(t)D_2(t) = D_2(t)\frac{d}{dt}\left(\frac{\Gamma(t)}{D_2(t)}\right) - \Gamma^2(t)$$

$$-\frac{d}{dt}\left(\frac{W(R_2,D_2)}{R_2 D_2}\right) + \left(2\Gamma(t) + \frac{d}{dt}\ln R_2(t)\right)\frac{W(R_2,D_2)}{R_2 D_2} \tag{29}$$

$$= D_2(t)\frac{d}{dt}\left(\frac{\Gamma(t)}{D_2(t)}\right) - \Gamma^2(t) + \left(2\Gamma(t) + \frac{d}{dt}\ln R_2(t)\right)\frac{d}{dt}\ln\frac{D_2(t)}{R_2(t)} - \frac{d^2}{dt^2}\ln\frac{D_2(t)}{R_2(t)},$$

where $W(R_2,D) = R_2 D'_{2t} - D_2 R'_{2t}$ is the Wronskian.

After the substitutions

$$Q(x,t) = q(x,t) \exp\left[\int_0^t \Gamma(\tau)d\tau\right], \quad R(t) = R_2(t) \exp\left[2\int_0^t \Gamma(\tau)d\tau\right], \quad D(t) = D_2(t),$$

Eq. (28) is transformed to the generalized NLSE without gain or loss term

$$i\frac{\partial q}{\partial t} + \frac{1}{2}D(t)\frac{\partial^2 q}{\partial x^2} + \left[\sigma R(t)|q|^2 - 2\lambda_0(t)x - \frac{1}{2}\Omega^2(t)x^2\right]q = 0. \tag{30}$$

Finally, the Lax equation (2) with matrices (3-6) provides the nonautonomous model (30) under condition that dispersion $D(t)$, nonlinearity $R(t)$, and the harmonic potential satisfy to the following exact integrability conditions

$$\Omega^2(t)D(t) = \frac{W(R,D)}{RD}\frac{d}{dt}\ln R(t) - \frac{d}{dt}\left(\frac{W(R,D)}{RD}\right)$$

$$= \frac{d}{dt}\ln D(t)\frac{d}{dt}\ln R(t) - \frac{d^2}{dt^2}\ln D(t) - R(t)\frac{d^2}{dt^2}\frac{1}{R(t)}. \tag{31}$$

The self-induced soliton phase shift is given by

$$\Theta(t) = -\frac{W\left[(R(t),D(t)\right]}{D^2(t)R(t)} \tag{32}$$

and the time-dependent spectral parameter is represented by

$$\Lambda(t) = \kappa(t) + i\eta(t) = \frac{D_0 R(t)}{R_0 D(t)}\left[\Lambda(0) + \frac{R_0}{D_0}\int_0^t \frac{\lambda_0(\tau)D(\tau)}{R(\tau)}d\tau\right], \tag{33}$$

where the main parameters: time invariant eigenvalue $\Lambda(0) = \kappa_0 + i\eta_0$; $D_0 = D(0)$; $R_0 = R(0)$ are defined by the initial conditions.

We call Eq. (31) as the law of the soliton adaptation to the external potentials. The basic property of classical solitons to interact elastically holds true, but the novel feature of the nonautonomous solitons arises. Namely, both amplitudes and speeds of the solitons, and consequently, their spectra, during the propagation and after the interaction are no longer the same as those prior to the interaction. All nonautonomous solitons generally move with varying amplitudes $\eta(t)$ and speeds $\kappa(t)$ adapted both to the external potentials and to the dispersion $D(t)$ and nonlinearity $R(t)$ changes.

Having obtained the eigenvalue equations for scattering potential, we can write down the general solutions for bright ($\sigma = +1$) and dark ($\sigma = -1$) nonautonomous solitons applying the auto-Bäcklund transformation (Chen, 1974) and the recurrent relation

$$q_n(x,t) = -q_{n-1}(x,t) - \frac{4\eta_n\tilde{\Gamma}_{n-1}(x,t)}{1 + \left|\tilde{\Gamma}_{n-1}(x,t)\right|^2} \times \sqrt{\frac{D(t)}{R(t)}}\exp[-i\Theta x^2/2], \tag{34}$$

which connects the $(n-1)$ and n - soliton solutions by means of the so-called pseudo-potential $\tilde{\Gamma}_{n-1}(x,t) = \psi_1(x,t)/\psi_2(x,t)$ for the $(n-1)$−soliton scattering functions $\psi(x,t) = (\psi_1\psi_2)^T$.

Bright $q_1^+(x,t)$ and dark $q_1^-(x,t)$ soliton solutions are represented by the following analytic expressions:

$$q_1^+(x,t \mid \sigma = +1) = 2\eta_1(t)\sqrt{\frac{D(t)}{R(t)}}\,\text{sech}\,[\xi_1(x,t)] \times \exp\left\{-i\left(\frac{\Theta(t)}{2}x^2 + \chi_1(x,t)\right)\right\}; \quad (35)$$

$$q_1^-(x,t \mid \sigma = -1) = 2\eta_1(t)\sqrt{\frac{D(t)}{R(t)}}\left[\sqrt{(1-a^2)} + ia\tanh\zeta(x,t)\right] \quad (36)$$

$$\times \exp\left\{-i\left(\frac{\Theta(t)}{2}x^2 + \phi(x,t)\right)\right\},$$

$$\zeta(x,t) = 2a\eta_1(t)x + 4a\int_0^t D(\tau)\eta_1(\tau)\kappa_1(\tau)d\tau, \quad (37)$$

$$\phi(x,t) = 2\left[\kappa_1(t) - \eta_1(t)\sqrt{(1-a^2)}\right]x$$

$$+2\int_0^t D(\tau)\left[\kappa_1^2 + \eta_1^2\left(3-a^2\right) - 2\kappa_1\eta_1\sqrt{(1-a^2)}\right]d\tau. \quad (38)$$

Dark soliton (36) has an additional parameter, $0 \le a \le 1$, which designates the depth of modulation (the blackness of gray soliton) and its velocity against the background. When $a = 1$, dark soliton becomes black. For optical applications, Eq.(36) can be easily transformed into the Hasegawa and Tappert form for the nonautonomous dark solitons (Hasegawa, 1995) under the condition $\kappa_0 = \eta_0\sqrt{(1-a^2)}$ that corresponds to the special choice of the retarded frame associated with the group velocity of the soliton

$$q_1^-(x,t \mid \sigma = -1) = 2\eta_1(t)\sqrt{\frac{D(t)}{R(t)}}\left[\sqrt{(1-a^2)} + ia\tanh\tilde{\zeta}(x,t)\right]$$

$$\times \exp\left\{-i\left(\frac{\Theta(t)}{2}x^2 + \tilde{\phi}(x,t)\right)\right\},$$

$$\tilde{\zeta}(x,t) = 2a\eta_1(t)x + 4a\int_0^t D(\tau)\eta_1(\tau)\left[\eta_1(\tau)\sqrt{(1-a^2)} + K(\tau)\right]d\tau,$$

$$\tilde{\phi}(x,t) = 2K(t)x + 2\int_0^t D(\tau)\left[K^2(\tau) + 2\eta_1^2(\tau)\right]d\tau,$$

$$K(t) = \frac{R(t)}{D(t)}\int_0^t \lambda_0(\tau)\frac{D(\tau)}{R(\tau)}d\tau.$$

Notice that the solutions considered here hold only when the nonlinearity, dispersion and confining harmonic potential are related by Eq. (31), and both $D(t) \ne 0$ and $R(t) \ne 0$ for all times by definition.

Two-soliton $q_2(x,t)$ solution for $\sigma = +1$ follows from Eq. (34)

$$q_2(x,t) = 4\sqrt{\frac{D(t)}{R(t)} \frac{N(x,t)}{D(x,t)}} \exp\left[-\frac{i}{2}\Theta(t)x^2\right],$$ (39)

where the numerator $N(x,t)$ is given by

$$N = \cosh\tilde{\xi}_2 \exp(-i\chi_1)$$

$$\times [(\kappa_2 - \kappa_1)^2 + 2i\eta_2(\kappa_2 - \kappa_1)\tanh\tilde{\xi}_2 + \eta_1^2 - \eta_2^2] + \eta_2 \cosh\tilde{\xi}_1 \exp(-i\chi_2)$$

$$\times [(\kappa_2 - \kappa_1)^2 - 2i\eta_1(\kappa_2 - \kappa_1)\tanh\tilde{\xi}_1 - \eta_1^2 + \eta_2^2],$$ (40)

and the denominator $D(x,t)$ is represented by

$$D = \cosh(\tilde{\xi}_1 + \tilde{\xi}_2)\left[(\kappa_2 - \kappa_1)^2 + (\eta_2 - \eta_1)^2\right]$$

$$+ \cosh(\tilde{\xi}_1 - \tilde{\xi}_2)\left[(\kappa_2 - \kappa_1)^2 + (\eta_2 + \eta_1)^2\right] - 4\eta_1\eta_2 \cos(\chi_2 - \chi_1).$$ (41)

Arguments and phases in Eqs.(39-41)

$$\tilde{\xi}_i(x,t) = 2\eta_i(t)x + 4\int_0^t D(\tau)\eta_i(\tau)\kappa_i(\tau)d\tau,$$ (42)

$$\chi_i(x,t) = 2\kappa_i(t)x + 2\int_0^t D(\tau)\left[\kappa_i^2(\tau) - \eta_i^2(\tau)\right]d\tau$$ (43)

are related with the amplitudes

$$\eta_i(t) = \frac{D_0 R(t)}{R_0 D(t)}\eta_{0i},$$ (44)

and velocities

$$\kappa_i(t) = \frac{D_0 R(t)}{R_0 D(t)}\left[\kappa_{0i} + \frac{R_0}{D_0}\int_0^t \frac{\lambda_0(\tau)D(\tau)}{R(\iota)}d\tau\right]$$ (45)

of the nonautonomous solitons, where κ_{0i} and η_{0i} correspond to the initial velocity and amplitude of the i-th soliton ($i = 1, 2$).

Eqs. (39-45) describe the dynamics of two bounded solitons at all times and all locations. Obviously, these soliton solutions reduce to classical soliton solutions in the limit of autonomous nonlinear and dispersive systems given by conditions: $R(t) = D(t) = 1$, and $\lambda_0(t) = \Omega(t) \equiv 0$ for canonical NLSE without external potentials.

5. Chirped optical solitons with moving spectra in nonautonomous systems: colored nonautonomous solitons

Both the nonlinear Schrödinger equations (28, 30) and the Lax pair equations (3–6) are written down here in the most general form. The transition to the problems of optical solitons is accomplished by the substitution $x \to T$ (or $x \to X$); $t \to Z$ and $q^+(x,t) \to \tilde{u}^+(Z,T(\,or\,X))$ for bright solitons, and $[q^-(x,t)]^* \to \tilde{u}^-(Z,T(\,or\,X))$ for dark solitons, where the asterisk denotes the complex conjugate, Z is the normalized distance, and T is the retarded time for temporal solitons, while X is the transverse coordinate for spatial solitons.

The important special case of Eq.(30) arises under the condition $\Omega^2(Z) = 0$. Let us rewrite Eq. (30) by using the reduction $\Omega = 0$, which denotes that the confining harmonic potential is vanishing

$$i\frac{\partial u}{\partial Z} + \frac{\sigma}{2}D(Z)\frac{\partial^2 u}{\partial T^2} + R(Z)\,|u|^2\,u - 2\sigma\lambda_0(Z)Tu = 0. \tag{46}$$

This implies that the self-induced soliton phase shift $\Theta(Z)$, dispersion $D(Z)$, and nonlinearity $R(Z)$ are related by the following law of soliton adaptation to external linear potential

$$D(Z)/D_0 = R(Z)/R_0 \exp\left\{ -\frac{\Theta_0 D_0}{R_0} \int\limits_0^Z R(\tau)d\tau \right\}. \tag{47}$$

Nonautonomous exactly integrable NLSE model given by Eqs. (46,47) can be considered as the generalization of the well-studied Chen and Liu model (Chen, 1976) with linear potential $\lambda_0(Z) \equiv \alpha_0 = const$ and $D(Z) = D_0 = R(Z) = R_0 = 1$, $\sigma = +1$, $\Theta_0 = 0$. It is interesting to note that the accelerated solitons predicted by Chen and Liu in plasma have been discovered in nonlinear fiber optics only decade later (Agrawal, 2001; Dianov et al., 1989; Taylor, 1992). Notice that nonautonomous solitons with nontrivial self-induced phase shifts and varying amplitudes, speeds and spectra for Eq. (46) are given in quadratures by Eqs. (35-45) under condition $\Omega^2(Z) = 0$.

Let us show that the so-called Raman colored optical solitons can be approximated by this equation. Self-induced Raman effect (also called as soliton self-frequency shift) is being described by an additional term in the NLSE: $-\sigma_R U\partial\mid U\mid^2/\partial T$, where σ_R originates from the frequency dependent Raman gain (Agrawal, 2001; Dianov et al., 1989; Taylor, 1992). Assuming that soliton amplitude does not vary significantly during self-scattering $\mid U\mid^2 = \eta^2 sech^2(\eta T)$, we obtain that

$$\sigma_R\frac{\partial\mid U\mid^2}{\partial T} \approx -2\sigma_R\eta^4 T = 2\alpha_0 T$$

and $dv/dZ = \sigma_R\eta^4/2$, where $v = \kappa/2$. The result of soliton perturbation theory (Agrawal, 2001; Dianov et al., 1989; Taylor, 1992) gives $dv/dZ = 8\sigma_R\eta^4/15$. This fact explains the remarkable stability of colored Raman solitons that is guaranteed by the property of the exact integrability of the Chen and Liu model (Chen, 1976). More general model Eq. (46) and its exact soliton solutions open the possibility of designing an effective soliton compressor, for example, by drawing a fiber with $R(Z) = 1$ and $D(Z) = \exp(-c_0 Z)$, where $c_0 = \Theta_0 D_0$. It seems very attractive to use the results of nonautonomous solitons concept in ultrashort photonic applications and soliton lasers design.

Another interesting feature of the novel solitons, which we called colored nonautonomous solitons, is associated with the nontrivial dynamics of their spectra. Frequency spectrum of the chirped nonautonomous optical soliton moves in the frequency domain. In particular, if dispersion and nonlinearity evolve in unison $D(t) = R(t)$ or $D = R = 1$, the solitons propagate with identical spectra, but with totally different time-space behavior.

Consider in more details the case when the nonlinearity $R = R_0$ stays constant but the dispersion varies exponentially along the propagation distance

$$D(Z) = D_0 \exp(-c_0 Z),$$
$$\Theta(Z) = \Theta_0 \exp(c_0 Z).$$

Let us write the one and two soliton solutions in this case with the lineal potential that, for simplicity, does not depend on time: $\lambda_0(Z) = \alpha_0 = const$

$$U_1(Z,T) = 2\eta_{01}\sqrt{D_0 \exp(c_0 Z)}\operatorname{sech}\left[\xi_1(Z,T)\right] \times \exp\left[-\frac{i}{2}\Theta_0 \exp(c_0 Z) T^2 - i\chi_1(Z,T)\right], \quad (48)$$

$$U_2(Z,T) = 4\sqrt{D_0 \exp(-c_0 Z)}\frac{N(Z,T)}{D(Z,T)}\exp\left[-\frac{i}{2}\Theta_0 \exp(c_0 Z) T^2\right], \quad (49)$$

where the nominator $N(Z,T)$ and denominator $D(Z,T)$ are given by Eqs. (40,41) and

$$\xi_i(Z,T) = 2\eta_{0i}T \exp(c_0 Z) + 4D_0\eta_{0i}$$

$$\times \left\{\frac{\kappa_{0i}}{c_0}\left[\exp(c_0 Z) - 1\right] + \frac{\alpha_0}{c_0}\left[\frac{\exp(c_0 Z) - 1}{c_0} - Z\right]\right\}, \quad (50)$$

$$\chi_i(Z,T) = 2\kappa_{0i}T \exp(c_0 Z) + 2D_0\left(\kappa_{0i}^2 - \eta_{0i}^2\right)\frac{\exp(2c_0 Z) - 1}{2c_0}$$

$$+2T\frac{\alpha_0}{c_0}\left[\exp(c_0 Z) - 1\right] + 4D_0\kappa_{0i}\frac{\alpha_0}{c_0}\left[\frac{\exp(c_0 Z) - 1}{c_0} - t\right]$$

$$+2D_0\left(\frac{\alpha_0}{c_0}\right)^2\left[\frac{\exp(c_0 Z) - \exp(-c_0 Z)}{c_0} - 2Z\right]. \quad (51)$$

The initial velocity and amplitude of the i-th soliton ($i = 1,2$) are denoted by κ_{0i} and η_{0i}.

We display in Fig.1(a,b) the main features of nonautonomous colored solitons to show not only their acceleration and reflection from the lineal potential, but also their compression and amplitude amplification. Dark soliton propagation and dynamics are presented in Fig.1(c,d). The limit case of the Eqs.(48-51) appears when $c_0 \to \infty$ (that means $D(Z) = D_0$ =constant) and corresponds to the Chen and Liu model (Chen, 1976). The solitons with argument and phase

$$\xi(Z,T) = 2\eta_0\left(T + 2\kappa_0 Z + \alpha_0 Z^2 - T_0\right),$$

$$\chi(Z,T) = 2\kappa_0 T + 2\alpha_0 TZ + 2\left(\kappa_0^2 - \eta_0^2\right)Z + 2\kappa_0\alpha_0 Z^2 + \frac{2}{3}\alpha_0^2 Z^3$$

represents the particle-like solutions which may be accelerated and reflected from the lineal potential.

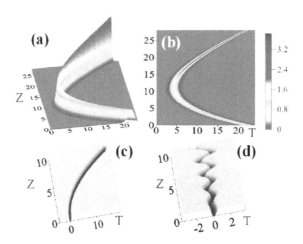

Fig. 1. Evolution of nonautonomous bright (a,b) optical soliton calculated within the framework of the generalized model given by Eqs. (46-51) after choosing the soliton management parameters c_0=0.05, $\alpha_0 = -0.2$, $\eta_{10} = 0.5$, $\kappa_{10} = 1.5$. (a) the temporal behavior; (b) the corresponding contour map. (c,d) Dark nonautonomous soliton dynamics within the framework of the model Eqs. (46,47) after choosing the soliton management parameters: (c) R=–D=1.0 and $\alpha_0 = -1.0$ and (d) R=–D=cos(ωZ), where $\omega = 3.0$.

6. Bound states of colored nonautonomous optical solitons: nonautonomous "agitated" breathers.

Let us now give the explicit formula of the soliton solutions (48,49) for the case where all eigenvalues are pure imaginary, or the initial velocities of the solitons are equal to zero. In the case $N = 1$ and $\lambda_0(Z) = 0$, we obtain

$$U_1(Z,T) = 2\eta_{01}\sqrt{D_0 \exp(c_0 Z)}\,\text{sech}\left[2\eta_{01}T \exp(c_0 Z)\right]$$
$$\times \exp\left[-\frac{i}{2}\Theta_0 \exp(c_0 Z)\,T^2 + i2D_0\eta_{01}^2\frac{\exp(2c_0 Z)-1}{2c_0}\right]. \qquad (52)$$

This result shows that the laws of soliton adaptation to the external potentials (31) allow to stabilize the soliton even without a trapping potential. In addition, Eq.(52) indicates the possibility for the optimal compression of solitons, which is shown in Fig.2. We stress that direct computer experiment confirms the exponential in time soliton compression scenario in full accordance with analytical expression Eq.(52).

The bound two-soliton solution for the case of the pure imaginary eigenvalues is represented by

$$U_2(Z,T) = 4\sqrt{D_0 \exp(-c_0 Z)}\frac{N(Z,T)}{D(Z,T)}\exp\left[-\frac{i}{2}\Theta_0 \exp(c_0 Z)\,T^2\right], \qquad (53)$$

where

$$N = \left(\eta_{01}^2 - \eta_{02}^2\right)\exp(c_0 Z)\left[\eta_{01}\cosh\xi_2 \exp(-i\chi_1) - \eta_{02}\cosh\xi_1 \exp(-i\chi_2)\right], \qquad (54)$$

$$D = \cosh(\xi_1 + \xi_2)\left(\eta_{01} - \eta_{02}\right)^2 + \cosh(\xi_1 - \xi_2)\left(\eta_{01} + \eta_{02}\right)^2 - 4\eta_{01}\eta_{02}\cos(\chi_2 - \chi_1), \qquad (55)$$

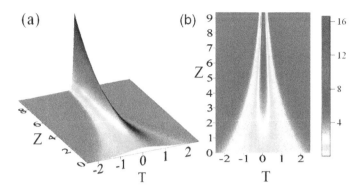

Fig. 2. Self-compression of nonautonomous soliton calculated within the framework of the model Eq. (46) after choosing the soliton management parameters $c_0 = 0.05; \alpha = 0$ and $\eta_0 = 0.5$. (a) the temporal behavior; (b) the corresponding contour map.

and

$$\xi_i(Z,T) = 2\eta_{0i}T \exp(c_0 Z), \tag{56}$$

$$\chi_i(Z,T) = -2D_0\eta_{0i}^2 \frac{\exp(2c_0 Z) - 1}{2c_0} + \chi_{i0}. \tag{57}$$

For the particular case of $\eta_{10} = 1/2$, $\eta_{20} = 3/2$ Eqs.(53-57) are transformed to

$$U_2(Z,T) = 4\sqrt{D_0 \exp(-c_0 Z)} \exp\left[-\frac{i}{2}\Theta_0 \exp(c_0 Z) T^2\right] \tag{58}$$

$$\times \exp\left[\frac{i}{4c_0}D_0\left[\exp(2c_0 Z) - 1\right] + \chi_{10}\right]$$

$$\times \frac{\cosh 3X - 3\cosh X \exp\{i2D_0\left[\exp(2c_0 Z) - 1\right]/c_0 + i\Delta\varphi\}}{\cosh 4X + 4\cosh 2X - 3\cos\{2D_0\left[\exp(2c_0 Z) - 1\right]/c_0 + \Delta\varphi\}},$$

where $X = T\exp(c_0 Z)$, $\Delta\varphi = \chi_{20} - \chi_{10}$.

In the $D(Z) = D_0 = 1$, $c_0 = 0$ limit, this solution is reduced to the well-known breather solution, which was found by Satsuma and Yajima (Satsuma & Yajima, 1974) and was called as the Satsuma-Yajima breather:

$$U_2(Z,T) = 4\frac{\cosh 3T + 3\cosh T \exp(4iZ)}{\cosh 4T + 4\cosh 2T + 3\cos 4Z}\exp\left(\frac{iZ}{2}\right).$$

At $Z = 0$ it takes the simple form $U(Z,T) = 2sech(T)$. An interesting property of this solution is that its form oscillates with the so-called soliton period $T_{sol} = \pi/2$.

In more general case of the varying dispersion, $D(Z) = D_0 \exp(-c_0 Z)$, shown in Fig.3 ($c_0 = 0.25$, $\eta_{10} = 0.25$, $\eta_{20} = 0.75$), the soliton period, according to Eq.(58), depends on time. The Satsuma and Yajima breather solution can be obtained from the general solution if and only if the soliton phases are chosen properly, precisely when $\Delta\varphi = \pi$. The intensity profiles of the wave build up a complex landscape of peaks and valleys and reach their peaks at the points of the maximum. Decreasing group velocity dispersion (or increasing nonlinearity) stimulates the Satsuma-Yajima breather to accelerate its period of "breathing" and to increase its peak amplitudes of "breathing", that is why we call this effect as "agitated breather" in nonautonomous system.

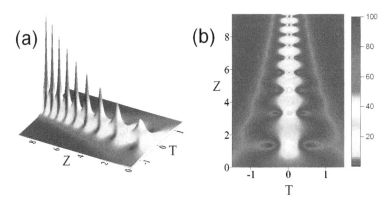

Fig. 3. Nonautonomous "agitated" breather (58) calculated within the framework of the model (46) after choosing the soliton management parameters $c_0 = 0.25$, $\eta_{10} = 0.5$, $\eta_{20} = 1.5$. (a) the temporal behavior; (b) the corresponding contour map.

7. Rogue waves, "quantized" modulation instability, and dynamics of nonautonomous Peregrine solitons under "hyperbolic hurricane wind"

Recently, a method of producing optical rogue waves, which are a physical counterpart to the rogue (monster) waves in oceans, have been developed (Solli et al., 2007). Optical rogue waves have been formed in the so-called soliton supercontinuum generation, a nonlinear optical process in which broadband "colored" solitons are generated from a narrowband optical background due to induced modulation instability and soliton fission effects (Dudley, 2009; Dudley et al., 2006; 2008).

Ordinary, the study of rogue waves has been focused on hydrodynamic applications and experiments (Clamond et al., 2006; Kharif & Pelinovsky, 2003). Nonlinear phenomena in optical fibers also support rogue waves that are considered as soliton supercontinuum noise. It should be noticed that because optical rogue waves are closely related to oceanic rogue waves, the study of their properties opens novel possibilities to predict the dynamics of oceanic rogue waves. By using the mathematical equivalence between the propagation of nonlinear waves on water and the evolution of intense light pulses in optical fibers, an international research team (Kibler et al., 2010) recently reported the first observation of the so-called Peregrine soliton (Peregrine, 1983). Similar to giant nonlinear water waves, the Peregrine soliton solutions of the NLSE experience extremely rapid growth followed by just as rapid decay (Peregrine, 1983). Now, the Peregrine soliton is considered as a prototype of the famous ocean monster (rogue) waves responsible for many maritime catastrophes.

In this Section, the main attention will be focused on the possibilities of generation and amplification of nonautonomous Peregrine solitons. This study is an especially important for understanding how high intensity rogue waves may form in the very noisy and imperfect environment of the open ocean.

First of all, let us summarize the main features of the phenomenon known as the induced modulation instability. In 1984, Akira Hasegawa discovered that modulation instability of continuous (cw) wave optical signal in a glass fiber combined with an externally applied amplitude modulation can be utilized to produce a train of optical solitons (Hasegawa,

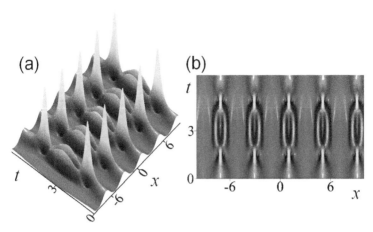

Fig. 4. Illustrative example of the temporal-spatial dynamics of the induced modulation instability and the Fermi-Pasta-Ulam recurrence effect calculated in the framework of the canonical NLSE model : (a) the intensity distribution; (b) the corresponding contour map.

1984). In the sense that the external modulation induces the modulation instability, Hasegawa called the total process as the induced modulation instability. To demonstrate the induced modulation instability (IMI), following Hasegawa, we solved the NLSE numerically with different depths and wavelength of modulation of cw wave. The main features of the induced modulation instability are presented in Fig.4. In Figure 4, following Hasegawa (Hasegawa, 1984), we present the total scenario of IMI and the restoration of the initial signal due to the Fermi-Pasta-Ulama recurrence effect. In our computer experiments, we have found novel and interesting feature of the IMI. Varying the depth of modulation and the level of continuous wave, we have discovered the effect which we called a "quantized" IMI. Figure 5 shows typical results of the computation. As can be clearly seen, the high-intensity IMI peaks are formed and split periodically into two, three, four, and more high-intensity peaks. In Fig.5 we present this splitting ("quantization") effect of the initially sinus like modulated cw signal into two and five high-intensity and "long-lived" components.

The Peregrine soliton can be considered as the utmost stage of the induced modulation instability, and its computer simulation is presented in Fig.6 When we compare the high-energy peaks of the IMI generated upon a distorted background (see Figs.4, 5) with exact form of the Peregrine soliton shown in Fig.7(a) we can understand, how such extreme wave structures may appear as they emerge suddenly on an irregular surface such as the open ocean.

There are two basic questions to be answered. What happens if arbitrary modulated cw wave is subjected to some form of external force? Such situations could include effects of wind, propagation of waves in nonuniform media with time dependent density gradients and slowly varying depth, nonlinearity and dispersion. For example, in Fig.7(b), we show the possibility of amplification of the Peregrine soliton when effects of wind are simulated by additional gain term in the canonical NLSE. The general questions naturally arise: To what extent the Peregrine soliton can be amplified under effects of wind, density gradients and

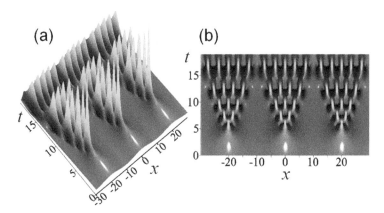

Fig. 5. Illustrative example of the "quantized" induced modulation instability: (a) the temporal-spatial behavior; (b) the corresponding contour map.

slowly varying depth, nonlinearity and dispersion? To answer these questions, let us consider the dynamics of the Peregrine soliton in the framework of the nonautonomous NLSE model. In the previous chapters, the auto -Bäcklund transformation has been used to find soliton solutions of the nonautonomous NLSE model. Now, we consider another remarkable method to study nonautonomous solitons. The following transformation

$$q(x,t) = A(t)u(X,T)\exp\left[i\phi(X,T)\right] \tag{59}$$

has been used by Serkin and Hasegawa in (Serkin & Hasegawa, 2000a;b; 2002) to reduce the nonautonomous NLSE with varying dispersion, nonlinearity and gain or loss to the "ideal" NLSE

$$i\frac{\partial u}{\partial T} + \frac{\sigma}{2}\frac{\partial^2 u}{\partial X^2} + |u|^2 u = 0,$$

where the following notations may be introduced

$$A(t) = \sqrt{P(t)}; \quad X = P(t)x; \quad T(t) = \int_0^t D(\tau)P^2(\tau)d\tau; \tag{60}$$

$$\phi(X,T) = \frac{1}{2}\frac{W(R,D)}{R^3}X^2 - \varphi(X,T), \tag{61}$$

where $\varphi(X,T)$ is the phase of the canonical soliton.

It is easy to see that by using Eq.(59-61), the one-soliton solution may be written in the following form

$$q_1^+(x,t \mid \sigma = +1) = 2\widetilde{\eta}_0 A(t)\text{sech}\left[2\widetilde{\eta}_0 X + 4\widetilde{\eta}_0\widetilde{\kappa}_0 T(t)\right]$$

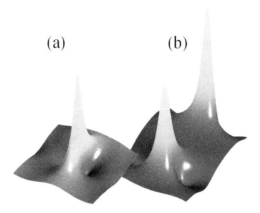

Fig. 6. Illustrative examples of the Peregrine soliton dynamics: (a) - classical Peregrine soliton calculated in the framework of the canonical NLSE model; (b) its behavior under linear amplification associated with continuous wind.

$$\times \exp\left\{i\left[\frac{1}{2}\frac{W(R,D)}{R^3}X^2 - 2\tilde{\kappa}_0 X - 2(\tilde{\kappa}_0^2 - \tilde{\eta}_0^2)T(t)\right]\right\},$$

$$\tilde{\eta}_0 = \frac{D_0}{R_0}\eta_0; \quad \tilde{\kappa}_0 = \frac{D_0}{R_0}\kappa_0; \quad P(t) = R(t)/D(t). \tag{62}$$

The transformation (59) can be applied to obtain all solutions of the nonautonomous NLSE (30) and, in particular, the nonautonomous rational solutions known as the Peregrine solitons. Thus, the Peregrine soliton (Peregrine, 1983) can be discovered for the nonautonomous NLSE model as well

$$q_P(x,t) = A(t)r(X,T)\exp\left[i\phi(T)\right] \tag{63}$$

where

$$r(X,T) = 1 - \frac{4(1+2iT)}{1+4T^2+4X^2}, \tag{64}$$

$$\phi(X,T) = \frac{1}{2}\frac{W(R,D)}{R^3}X^2 + T(t) \tag{65}$$

Figure 7 shows spatiotemporal behavior of the nonautonomous Peregrine soliton. The nonautonomous Peregrine soliton (63-65) shown in Fig.7(b) has been calculated in the framework of the nonautonomous NLSE model (28) after choosing the parameters $\lambda_0 = \Omega = 0, D_2 = R_2 = 1$ and the gain coefficient $\Gamma(t) = \Gamma_0/(1-\Gamma_0 t)$. Somewhat surprisingly, however, this figure indicates a sharp compression and strong amplification of the nonautonomous Peregrine soliton under the action of hyperbolic gain which, in particular, in the open ocean can be associated with "hyperbolic hurricane wind".

It should be stressed that since the nonautonomous NLSE model is applied in many other physical systems such as plasmas and Bose-Einstein condensates (BEC), the results obtained in this Section can stimulate new research directions in many novel fields (see, for example, (Bludov et al., 2009; Yan, 2010)).

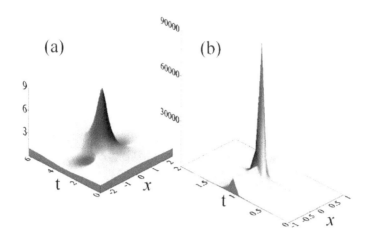

Fig. 7. (a) Autonomous and (b) nonautonomous Peregrine solitons calculated within the framework of the model (63-65) after choosing the soliton management parameters $\Gamma_0 = 0.33$.

8. Nonautonomous KdV solitons

Notice, that the nonlinear evolution equations that arise in the approach of variable spectral parameter contain, as a rule, an explicit dependence on the coordinates. Our general approach makes it possible to construct not only the well-known equations, but also a number of new integrable equations (NLSE, KdV, modified KdV, Hirota and Satsuma and so on) by extending the Zakharov–Shabat (ZS) and AKNS formalism. In particular, Eqs.(9,10) under the conditions (11) with $a_2=0$, $a_3=-4iD_3$ and $R=1$ become

$$Q_T = -D_3 Q_{SSS} S_x^3 - 6iD_3 Q_{SS}\varphi_S S_x^3 + 3iD_3\sigma F^{2\gamma}Q^2\varphi_S S_x + 6D_3\sigma F^{2\gamma}QQ_S S_x \tag{66}$$

$$+Q_S\left(-S_t + \lambda_1 S - V_1 S_x - 6iD_3\varphi_{SS}S_x^3 + \frac{3}{4}D_3\varphi_S^2 S_x^3\right)$$

$$-iQ\left[2\lambda_0 S/S_x - 2\gamma + \frac{1}{2}(\varphi_T + \varphi_S S_t) - \frac{1}{2}\lambda_1 S\varphi_S + \frac{1}{2}V\varphi_S S_x\right]$$

$$+Q\left(\lambda_1 - \gamma\frac{F_T}{F} + \frac{3}{4}D_3\varphi_S\varphi_{SS}S_x^3\right) - iQ\left(-\frac{1}{8}D_3\varphi_S^3 S_x^3 + \frac{1}{2}D_3\varphi_{SSS}S_x^3\right),$$

Eq.(66) can be rewritten in the independent variables (x, t)

$$Q_t = -D_3 Q_{xxx} - 6iD_3 Q_{xx}\varphi_x + 3iD_3\sigma F^{2\gamma}Q^2\varphi_x + 6D_3\sigma F^{2\gamma}QQ_x \tag{67}$$

$$+Q_x\left(\lambda_1 S/S_x - V_1 - 6iD_3\varphi_{xx} + \frac{3}{4}D_3\varphi_x^2\right)$$

$$-iQ\left[2\lambda_0 S/S_x - 2\gamma + \frac{1}{2}(\varphi_T + \varphi_S S_t) - \frac{1}{2}\lambda_1 S\varphi_x/S_x + \frac{1}{2}V\varphi_x\right]$$

$$+Q\left(\lambda_1 - \gamma\frac{F_t}{F} + \frac{3}{4}D_3\varphi_x\varphi_{xx}\right) - iQ\left(-\frac{1}{8}D_3\varphi_x^3 + \frac{1}{2}D_3\varphi_{xxx}\right).$$

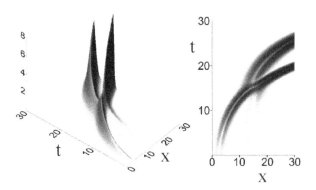

Fig. 8. Nonautonomous KdV solitons calculated within the framework of the model (71) after choosing the soliton management parameters $\alpha = 0.15$, $\eta_{10} = 0.40$, $\eta_{20} = 0.75$. On the left hand side the temporal behavior is presented, while the corresponding contour map is presented on the right hand side.

Let us consider the simplest option to choose the real solution $Q(x, t)$, which leads to the only possibility of $\varphi = \lambda_1 = 0$. In this case, Eq.(67) is reduced to the KdV with variable coefficients

$$Q_t - 6\sigma R_3(t) Q Q_x + D_3(t) Q_{xxx} + \frac{1}{2} \frac{W(D_3, R_3)}{D_3 R_3} = 0, \tag{68}$$

where the notation $R_3(t) = F^{2\gamma} D_3(t)$ has been introduced. It is easy to verify that Eq.(68) can be mapped into the standard KdV under the transformations

$$Q(x, t) = \frac{D_3(T)}{R_3(T)} q(x, T),$$

where $T = \int\limits_0^t D_3(\tau) d\tau$ so that $q(x, T)$ is given by the canonical KdV:

$$q_t - 6\sigma q q_x + q_{xxx} = 0.$$

Applying the auto-Backlund transformation, we can write down the two-soliton solution of the nonautonomous KdV

$$Q_2(x, t) = -2\sigma(\beta_1 - \beta_2) \frac{D_3(T)}{R_3(T)} \frac{\mathfrak{N}_1}{\mathfrak{D}_1}, \tag{69}$$

where

$$\mathfrak{N}_1 = \beta_1 \left(\sinh \xi_2\right)^2 + \beta_2 \left(\cosh \xi_1\right)^2, \tag{70}$$

$$\mathfrak{D}_1 = \left[\sqrt{2\beta_1} \sinh \xi_1 \sinh \xi_2 - \sqrt{2\beta_2} \cosh \xi_1 \cosh \xi_2\right]^2,$$

$$\xi_i = \sqrt{\beta_i/2} \left(x - 2\beta_i T\right), \ \beta_i = 2\eta_{0i}^2, \ i = 1, 2;$$

and $\eta_{02} > \eta_{01}$ are initial amplitudes of the solitons.

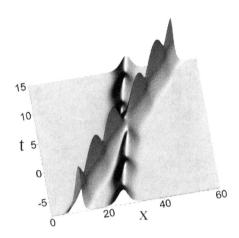

Fig. 9. Nonautonomous KdV solitons calculated within the framework of the model (72) after choosing the soliton management parameters $\alpha = 2.0$, $\beta = -0.25$, $\eta_{10} = 0.40$, $\eta_{20} = 0.75$.

As two illustrative examples, in Fig.8, we present the behavior of nonautonomous KdV soliton in the framework of the model

$$Q_t - 6\sigma QQ_x + \exp(\alpha t)Q_{xxx} - \frac{1}{2}\alpha Q = 0 \tag{71}$$

with lineal gain (or loss) accompanying by exponential variation of the dispersion coefficient; and in Fig.9 we show the dynamics of the KdV soliton in the nonautonomous system described by the model

$$Q_t - 6\sigma QQ_x + [1 + \beta\cos(\alpha t)] / (1 + \beta)Q_{xxx} + \frac{\alpha\beta\sin(\alpha t)}{2[1 + \beta\cos(\alpha t)](1 + \beta)}Q = 0 \tag{72}$$

where $D_3(t) = [1 + \beta\cos(\alpha t)] / (1 + \beta)$, $R_3(t) = 1$.

It is important to compare our exactly integrable nonautonomous KdV model with the model proposed by Johnson to describe the KdV soliton dynamics under the influence of the depth variation (Johnson, 1997) and given by

$$u_X - 6\sigma D(X)^{-3/2}uu_\xi + D(X)^{1/2}u_{\xi\xi\xi} + \frac{1}{2}\frac{D_X}{D}u = 0. \tag{73}$$

We stress that after choosing the parameters $R_3(t) = \mathcal{D}(t)^{-3/2}$ and $D_3(t) = \mathcal{D}(t)^{1/2}$, the potential in Eq.(68) becomes $\frac{W(D_3, R_3)}{D_3 R_3} = -2\mathcal{D}'/\mathcal{D}$, which is very nearly similar to the potential in Eq.(73) calculated by Johnson (Johnson, 1997).

9. Conclusions

The solution technique based on the generalized Lax pair operator method opens the possibility to study in details the nonlinear dynamics of solitons in nonautonomous nonlinear and dispersive physical systems. We have focused on the situation in which the generalized nonautonomous NLSE model was found to be exactly integrable from the point of view of the inverse scattering transform method. We have derived the laws of a soliton adaptation to the external potential. It is precisely this soliton adaptation mechanism which was of prime physical interest in our Chapter. We clarified some examples in order to gain a better understanding into this physical mechanism which can be considered as the interplay between nontrivial time-dependent parabolic soliton phase and external time-dependent potential. We stress that this nontrivial time-space dependent phase profile of nonautonomous soliton depends on the Wronskian of nonlinearity $R(t)$ and dispersion $D(t)$ and this profile does not exist for canonical NLSE soliton when $R(t) = D(t) = 1$.

Several novel analytical solutions for water waves have been presented. In particular, we have found novel solutions for the generalized Peregrine solitons in inhomogeneous and nonautonomous systems, "quantized" modulation instability, and the exactly integrable model for the Peregrine solitons under "hyperbolic hurricane wind". It was shown that important mathematical analogies between optical rogue waves and the Peregrine solitons in water open the possibility to study optical rogue waves and water rogue waves in parallel and, due to the evident complexity of experiments with rogue waves in oceans, this method offers remarkable possibilities in studies nonlinear hydrodynamics problems by performing experiments in the nonlinear optical systems.

We would like to conclude by saying that the concept of adaptation is of primary importance in nature and nonautonomous solitons that interact elastically and generally move with varying amplitudes, speeds, and spectra adapted both to the external potentials and to the dispersion and nonlinearity changes can be fundamental objects of nonlinear science.

This investigation is a natural follow up of the works performed in collaboration with Professor Akira Hasegawa and the authors would like to thank him for this collaboration. We thank BUAP and CONACyT, Mexico for support.

10. References

Ablowitz, M. J., Kaup, D. J., Newell, A. C. & Segur, H. (1973). Nonlinear-evolution equations of physical significance, *Phys. Rev. Lett.* 31(2): 125–127.

Agrawal, G. P. (2001). *Nonlinear Fiber Optics, 3rd ed.*, Academic Press, San Diego.

Akhmediev, N. N. & Ankiewicz, A. (1997). *Solitons. Nonlinear pulses and beams*, Charman and Hall, London.

Akhmediev, N. N. & Ankiewicz, A. (2008). *Dissipative Solitons: From Optics to Biology and Medicine*, Springer-Verlag Berlin.

Atre, R., Panigrahi, P. K. & Agarwal, G. S. (2006). Class of solitary wave solutions of the one-dimensional gross-pitaevskii equation, *Phys. Rev. E* 73(5): 056611.

Avelar, A. T., Bazeia, D. & Cardoso, W. B. (2009). Solitons with cubic and quintic nonlinearities modulated in space and time, *Phys. Rev. E* 79(2): 025602.

Balakrishnan, R. (1985). Soliton propagation in nonuniform media, *Phys. Rev. A* 32(2): 1144–1149.

Belić, M., Petrović, N., Zhong, W.-P., Xie, R.-H. & Chen, G. (2008). Analytical light bullet solutions to the generalized $(3 + 1)$-dimensional nonlinear schrödinger equation, *Phys. Rev. Lett.* 101(12): 123904.

Belyaeva, T., Serkin, V., Agüero, M., Hernandez-Tenorio, C. & Kovachev, L. (2011). Hidden features of the soliton adaptation law to external potentials, *Laser Physics* 21: 258–263.

Bludov, Y. V., Konotop, V. V. & Akhmediev, N. (2009). Matter rogue waves, *Phys. Rev. A* 80(3): 033610.

Calogero, F. & Degasperis, A. (1976). Coupled nonlinear evolution equations solvable via the inverse spectral transform, and solitons that come back: the boomeron, *Lettere Al Nuovo Cimento* 16: 425–433.

Calogero, F. & Degasperis, A. (1982). *Spectral transform and solitons: tools to solve and investigate nonlinear evolution equations*, Elsevier Science Ltd.

Chen, H.-H. (1974). General derivation of bäcklund transformations from inverse scattering problems, *Phys. Rev. Lett.* 33(15): 925–928.

Chen, H. H. & Liu, C. S. (1976). Solitons in nonuniform media, *Phys. Rev. Lett.* 37(11): 693–697.

Chen, H. H. & Liu, C. S. (1978). Nonlinear wave and soliton propagation in media with arbitrary inhomogeneities, *Phys. Fluids* 21: 377–380.

Chen, S., Yang, Y. H., Yi, L., Lu, P. & Guo, D. S. (2007). Phase fluctuations of linearly chirped solitons in a noisy optical fiber channel with varying dispersion, nonlinearity, and gain, *Phys. Rev. E* 75(3): 036617.

Christiansen, P. L., Sorensen, M. P. & Scott, A. C. (2000). *Nonlinear science at the dawn of the 21st century*, Lecture Notes in Physics, Springer, Berlin.

Clamond, D., Francius, M., Grue, J. & Kharif, C. (2006). Long time interaction of envelope solitons and freak wave formations, *Eur. J. Mech. B/Fluids* 25(5): 536–553.

Dianov, E. M., Mamyshev, P. V., Prokhorov, A. M. & Serkin, V. N. (1989). *Nonlinear Effects in Optical Fibers*, Harwood Academic Publ., New York.

Dudley, J. M. & Taylor, J. R. (2009). Ten years of nonlinear optics in photonic crystal fibre, *Nature Photonics* 3: 85–90.

Dudley, J. M., Genty, G. & Coen, S. (2006). Supercontinuum generation in photonic crystal fiber, *Rev. Mod. Phys.* 78(4): 1135–1184.

Dudley, J. M., Genty, G. & Eggleton, B. J. (2008). Harnessing and control of optical rogue waves insupercontinuum generation, *Opt. Express* 16(6): 3644–3651.

Gardner, C. S., Greene, J. M., Kruskal, M. D. & Miura, R. M. (1967). Method for solving the korteweg-devries equation, *Phys. Rev. Lett.* 19(19): 1095–1097.

Gupta, M. R. & Ray, J. (1981). Extension of inverse scattering method to nonlinear evolution equation in nonuniform medium, *J. Math. Phys.* 22(10): 2180–2183.

Hao, R. & Zhou, G. (2008). Exact multi-soliton solutions in nonlinear optical systems, *Optics Communications* 281(17): 4474–4478.

Hasegawa, A. (1984). Generation of a train of soliton pulses by induced modulational instability in optical fibers, *Opt. Lett.* 9(7): 288–290.

Hasegawa, A. & Kodama, Y. (1995). *Solitons in Optical Communications*, Oxford University Press.

Hasegawa, A. & Matsumoto, M. (2003). *Optical Solitons in Fibers, 3rd Edition*, Springer-Verlag, Berlin.

He, X.-G., Zhao, D., Li, L. & Luo, H. G. (2009). Engineering integrable nonautonomous nonlinear schrödinger equations, *Phys. Rev. E* 79(5): 056610.

Hernandez, T. C., Villargan, V. E., Serkin, V. N., Aguero, G. M., Belyaeva, T. L., Pena, M. R. & Morales, L. L. (2005). Dynamics of solitons in the model of nonlinear schrödinger equation with an external harmonic potential: 1. bright solitons, *Quantum Electronics* 35(9): 778.

Hernandez-Tenorio, C., Belyaeva, T. L. & Serkin, V. N. (2007). Parametric resonance for solitons in the nonlinear schrödinger equation model with time-dependent harmonic oscillator potential, *Physica B: Condensed Matter* 398(2): 460–463.

Herrera, J. J. E. (1984). Envelope solitons in inhomogeneous media, *J. Phys. A: Math. Gen.* 17(1): 95–98.

Johnson, R. S. (1997). *A modern introduction to the mathematical theory of water waves*, Cambridge University Press, New York.

Kharif, C. & Pelinovsky, E. (2003). Physical mechanisms of the rogue wave phenomenon, *Eur. J. Mech. B. Fluid* 22(6): 603–634.

Kharif, C., Pelinovsky, E. & Slunyaev, A. (2009). *Rogue Waves in the Ocean*, Springer-Verlag, Berlin.

Kibler, B., Fatome, J., Finot, C., Millot, G., Dias, F., Genty, G., Akhmediev, N. & Dudley, J. M. (2010). The peregrine soliton in nonlinear fibre optics, *Nature Physics* 6(10): 790–795.

Krumhansl, J. A. (1991). Unity in the science of physics, *Physics Today* 44(3): 33–38.

Lax, P. D. (1968). Integrals of nonlinear equations of evolution and solitary waves, *Commun. on Pure and Applied Mathematics* 21: 467–490.

Liu, W.-J., Tian, B. & Zhang, H.-Q. (2008). Types of solutions of the variable-coefficient nonlinear schrödinger equation with symbolic computation, *Phys. Rev. E* 78(6): 066613.

Luo, H., Zhao, D. & He, X. (2009). Exactly controllable transmission of nonautonomous optical solitons, *Phys. Rev. A* 79(6): 063802.

Nayfeh, A. H. & Balachandran, B. (2004). *Applied Nonlinear Dynamics*, Wiley-VCH Verlag GmbH & Co. KGaA, Weinheim.

Peregrine, D. H. (1983). Water waves, nonlinear schrödinger equations and their solutions, *Austral. Math. Soc. Ser. B* 25: 16–43.

Porsezian, K., Ganapathy, R., Hasegawa, A. & Serkin, V. (2009). Nonautonomous soliton dispersion management, *Quantum Electronics, IEEE Journal of* 45(12): 1577–1583.

Porsezian, K., Hasegawa, A., Serkin, V., Belyaeva, T. & Ganapathy, R. (2007). Dispersion and nonlinear management for femtosecond optical solitons, *Phys. Lett. A* 361(6): 504–508.

Satsuma, J. & Yajima, N. (1974). Initial value problems of one-dimensional self-modulation of nonlinear waves in dispersive media, *Prog. Theor. Phys. Supplement* 55: 284–306.

Serkin, V. N. & Belyaeva, T. L. (2001a). High-energy optical schrödinger solitons, *JETP Letters* 74: 573–577.

Serkin, V. N. & Belyaeva, T. L. (2001b). The lax representation in the problem of soliton management, *Quantum Electronics* 31(11): 1007.

Serkin, V. N. & Hasegawa, A. (2000a). Novel soliton solutions of the nonlinear schrödinger equation model, *Phys. Rev. Lett.* 85(21): 4502–4505.

Serkin, V. N. & Hasegawa, A. (2000b). Soliton management in the nonlinear schrödinger equation model with varying dispersion, nonlinearity, and gain, *JETP Letters* 72: 89–92.

Serkin, V. N. & Hasegawa, A. (2002). Exactly integrable nonlinear schrödinger equation models with varying dispersion, nonlinearity and gain: application for soliton dispersion, *Selected Topics in Quantum Electronics, IEEE Journal of* 8(3): 418–431.

Serkin, V. N., Hasegawa, A. & Belyaeva, T. L. (2004). Comment on "exact self-similar solutions of the generalized nonlinear schrödinger equation with distributed coefficients", *Phys. Rev. Lett.* 92(19): 199401.

Serkin, V. N., Hasegawa, A. & Belyaeva, T. L. (2007). Nonautonomous solitons in external potentials, *Phys. Rev. Lett.* 98(7): 074102.

Serkin, V. N., Hasegawa, A. & Belyaeva, T. L. (2010a). Nonautonomous matter-wave solitons near the feshbach resonance, *Phys. Rev. A* 81(2): 023610.

Serkin, V. N., Hasegawa, A. & Belyaeva, T. L. (2010b). Solitary waves in nonautonomous nonlinear and dispersive systems: nonautonomous solitons, *Journal of Modern Optics* 57(14-15): 1456–1472.

Serkin, V. N., Matsumoto, M. & Belyaeva, T. L. (2001a). Bright and dark solitary nonlinear bloch waves in dispersion managed fiber systems and soliton lasers, *Optics Communications* 196(1-6): 159–171.

Serkin, V. N., Matsumoto, M. & Belyaeva, T. L. (2001b). Nonlinear bloch waves, *JETP Letters* 73: 59–62.

Shin, H. J. (2008). Darboux invariants of integrable equations with variable spectral parameters, *Journal of Physics A: Mathematical and Theoretical* 41(28): 285201.

Solli, D. R., Ropers, C., Koonath, P. & Jalali, B. (2007). Optical rogue waves, *Nature* 450: 1054–1057.

Tappert, F. D. & Zabusky, N. J. (1971). Gradient-induced fission of solitons, *Phys. Rev. Lett.* 27(26): 1774–1776.

Taylor, J. R. (1992). *Optical solitons - theory and experiment*, Cambridge Univ. Press, Cambridge.

Tenorio, C. H., Villagran-Vargas, E., Serkin, V. N., Agüero-Granados, M., Belyaeva, T. L., Pena-Moreno, R. & Morales-Lara, L. (2005). Dynamics of solitons in the model of nonlinear schrödinger equation with an external harmonic potential: 2. dark solitons, *Quantum Electronics* 35(10): 929.

Wang, J., Li, L. & Jia, S. (2008). Nonlinear tunneling of optical similaritons in nonlinear waveguides, *J. Opt. Soc. Am. B* 25(8): 1254–1260.

Wu, L., Li, L. & Zhang, J. F. (2008). Controllable generation and propagation of asymptotic parabolic optical waves in graded-index waveguide amplifiers, *Phys. Rev. A* 78(1): 013838.

Wu, L., Zhang, J.-F., Li, L., Finot, C. & Porsezian, K. (2008). Similariton interactions in nonlinear graded-index waveguide amplifiers, *Phys. Rev. A* 78(5): 053807.

Yan, Z. (2010). Nonautonomous "rogons" in the inhomogeneous nonlinear schrödinger equation with variable coefficients, *Physics Letters A* 374(4): 672–679.

Zabusky, N. J. & Kruskal, M. D. (1965). Interaction of "solitons" in a collisionless plasma and the recurrence of initial states, *Phys. Rev. Lett.* 15(6): 240–243.

Zakharov, V. E. (1980). The inverse scattering method, *in* R. Bullough & P. J. Caudrey (eds), *Solitons*, Springer-Verlag, Berlin, pp. 243–285.

Zhang, J.-F., Wu, L. & Li, L. (2008). Self-similar parabolic pulses in optical fiber amplifiers with gain dispersion and gain saturation, *Phys. Rev. A* 78(5): 055801.

Zhao, D., He, X.-G. & Luo, H.-G. (2009). Transformation from the nonautonomous to standard nls equations, *Eur. Phys. J. D* 53: 213–216.

Zhao, D., Luo, H. G. & Chai, H. Y. (2008). Integrability of the gross-pitaevskii equation with feshbach resonance management, *Physics Letters A* 372(35): 5644–5650.

Part 2

Biological Applications and Biohydrodynamics

5

Laser-Induced Hydrodynamics in Water and Biotissues Nearby Optical Fiber Tip

V. I. Yusupov[1], V. M. Chudnovskii[1] and V. N. Bagratashvili[2]
*[1]V.I. Il'ichev Pacific Oceanological Institute, Far Eastern
Branch of Russian Academy of Sciences
[2]Institute of Laser and Information Technologies,
Russian Academy of Sciences
Russia*

1. Introduction

This paper is aimed at revealing the mechanisms of therapeutic effects stimulated by a medium power (1–10 W) fiber laser induced hydrodynamics in water-saturated bio-tissues. Modern laser medical technologies widely employ delivery of laser light to irradiated tissues via optical fibers. Optical fiber easily penetrates through needle and endoscopic channels, and laser light can be delivered through a fiber for puncture and endoscopic operations. Several laser medical technologies (puncture multichannel laser decompression of disc, laser intervention upon osteochondrosis, surgical treatment of chronic osteomyelitis, endovenous laser ablation, etc.) are based on effective hydrodynamic processes in water-saturated bio-tissues. These hydrodynamic processes trigger cellular response and regenerative effects through the specific mechanisms of mechano-biology. In this work, we consider different kinds of effects stimulated by a medium power laser-induced hydrodynamics in the vicinity of a fiber tip surface, in particular, generation of vapor-gas bubbles, fiber tip degradation, and generation of intense acoustic waves. Presence of strongly absorbed agents (in a form of Ag nanoparticles, in particular) in laser irradiated water nearby optical fiber tip results in appearance of pronounced filamentary structures of these agents.

2. Therapeutic motivation

One of the modern tendencies in a low-invasive medical therapy is a medium power (1–10 W) laser treatment of connective tissues. The examples of such technologies are: laser engineering of cartilages (Bagratashvili et al., 2006); puncture multichannel laser decompression of disc (Sandler et al., 2002; Sandler et al., 2004); laser intervention upon osteochondrosis (Chudnovskii & Yusupov, 2008); laser treatment of chronic osteomyelitis (Privalov et al., 2001); endovenous laser ablation (Van den Bos et al., 2009); fractional photothermolysis (Rokhsar & Ciocon, 2009).

Treatment of osteochondrosis, for example, is based on laser-induced (0.97 μm in wavelength and 2–10 W in power) formation of multiple channels inside an intervertebral disc using silica fiber with a carbon coated fiber tip surface, in order to enhance laser light absorption nearby the fiber tip. Osteochondrosis is caused by such partial destruction of intervertebral disc, followed by release of nucleus pulposus from disc in the form of hernia, which exerts pressure upon nervous roots thus giving pain. Fig 1a shows the scheme of formation of multiple laser channels inside intervertebral disc in the course of laser treatment of osteochondrosis (Sandler et al., 2002; Sandler et al., 2004; Chudnovskii & Yusupov, 2008). Transport laser delivery fiber passes inside the disc under treatment through a thin needle inserted to the disc (laser puncture procedure). Optical fiber is inserted through a thin needle via a posterolateral percutaneous approach under a local anesthesia. Important, that saline water is permanently introduced into the disc through the needle. Channel is formed by the heated fiber moving forward inside the disc. The fiber forms the channel and is shifted 1 -2 cm per 5 – 10 s inside the disc. Fig. 1b shows the example of such channels in nucleus pulposus of spinal disc formed by a fiber laser in the course of laboratory experiment (Sandler et al., 2004).

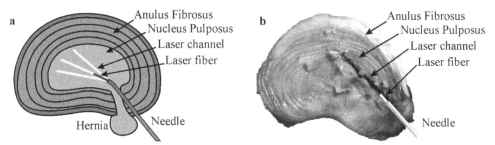

Fig. 1. a - Scheme of laser irradiation of spinal disc. b – Laser channel formed in spinal disc through optical fiber in presence of physiological solution (Sandler et al., 2004).

Surprisingly, that such action on herniated disc causes significant effect in some period of time on tissues located out of laser irradiated zone. As one can see, for example, on tomography picture (Fig. 2b), some cavities appear in the hernia, and its density decreases significantly compared with the density of hernia before laser treatment (Fig. 2a).

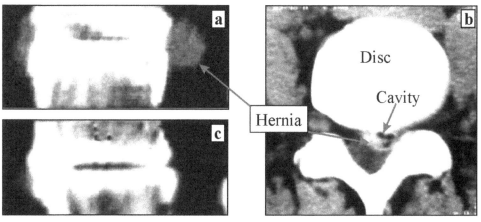

Fig. 2. Computer tomography pictures of herniated disc area. a– before healing: big sequester of hernia (side view); b - cavity inside hernia, stimulated by laser-induced channel formation in disc; c –three month after laser healing : no hernia.

As a result, the hernia transforms into a soft sponge, the pressure of hernia on nervous roots decreases, and relevant pain releases. The hernia itself disappears after some period of time, and regenerative processes take place which result (in a few month) in recovery of the disc structure and their main functions (Sandler et al., 2002, 2004; Chudnovskii et al., 2008, 2010a, 2010b).

Another important example of a medium power laser therapy is a laser treatment of chronic osteomyelitis. Fig. 3a demonstrates the X-ray image of femoral bone of the 14 year old patient heavily affected by osteomyelitis (Privalov et al., 2001). Significant destruction and rarefication of bone structure takes place. Typically, such bone tissue degradation requires amputation of organ. However, application of medium power laser treatment approach in this case gave, as a result, a complete regeneration of affected femoral bone (Fig. 3b), and no amputation was required. Again, therapy was based on a medium power laser-induced formation of channels (similar to that presented at Fig.1) in a bone medullary tissue, which stimulate successively the regeneration processes in the bone tissue.

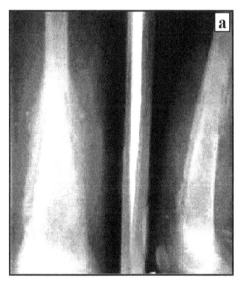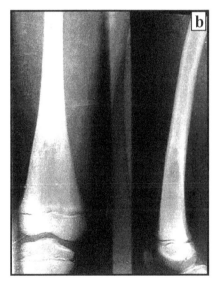

Fig. 3. X-ray images of right hip of the 14 y.o. patient with chronic osteomyelitis (Privalov et al., 2001). a - destruction and rarefication of bone structure before treatment. b - complete regeneration of bone structure 11 month after laser treatment.

Strong regenerative potential of medium power laser treatment for different kinds of tissues is already well recognized (Sandler et al., 2002, 2004; Chudnovskii & Yusupov, 2008, Chudnovskii et al., 2010a, 2010b), however the dominant primary physical mechanisms of such regeneration are still the subject of controversy. It is commonly accepted that the effects of medium power laser irradiation result from laser heating of tissues. However in most of cases, the pronounced therapeutic effect cannot be rationalized by laser-induced thermal tissue degradation only. For example, appearance of cavities in the hernia and significant decrease of its density observed immediately after laser manipulation (Fig. 2b), takes place without its heating, since hernia is located quite far from the area of laser-induced channel formation, and, thus, heating of hernia is negligible.

We believe that *effective hydrodynamic processes* play dominant role for the effect of a medium power laser-induced regeneration and healing of connective tissues diseases (intervertebral hernia, osteomyelitis and some other diseases) using laser puncture procedures (Chudnovskii & Yusupov, 2008; Chudnovskii et al., 2010a, 2010b). Main features of these processes will be considered below.

3. Laser-induced generation of micro-bubbles in water

The key process for the mechanism of medium power laser-induced regeneration and healing of musculoskeletal system diseases is the generation of micro-bubbles in inter-tissue water (Yusupov et al., 2010).

3.1 Laser-induced generation of micro-bubbles in a free water

Formation of micro-bubbles in a free water was studied with the aid of the optical methods using a water filled plastic cell (the horizontal dimensions are 150 × 100 mm and the height is 15 mm) and glass capillaries with an inner diameter of 1 mm. In the most of experiments, the working fiber tip is preliminary blackened by a short (~1 s) contact of the fiber tip with a wooden plate at a laser power of about 3 W. The fiber tip surface thus covers by a thin carbon layer owing to the wood burning. Such a procedure is well reproduced, so that from 10 to 20% of the laser power is absorbed in the thin carbon layer. Computer controlled fiber lasers (LS-0.97 and LS-1.55 of IRE–Polus, Russia) with the wavelengths of 0.97 μm and 1.55 μm, 1–10 W in power were interfaced with a 400 μm core diameter silica fiber. Low intensity (up to 1 mW) green pilot beam from the built in diode laser was used to highlight the laser irradiated zone in the cell. The fiber is horizontally fixed in the cell, which is placed on the worktable of a MICROS MC300 microscope equipped with a Vision digital color camera interfaced with PC. The water cell was also placed on the table with illumination, and the processes in the vicinity of the heated fiber tip were visualized using a Photron Fastcam SA-3 camera at rates of 2000 or 10000 frames per second. To control the laser induced spectrum, an Ocean Optics USB4000 fiber spectrum analyzer was used, which is interfaced with PC and has an optical resolution of about 1.5 nm and 200–1100 nm wavelength range. For better visualization of hydrodynamic flows the collargol (albumin coated Ag nanoparticles) have been added to water in the cell (Yusupov et al., 2011b).

Hydrodynamic flows taking place nearby the fiber tip when laser power is on, can be clearly seen in a scattering mode using illumination with green light of pilot laser beam through the same transport fiber (Fig. 4). Such flows result in intrusion of collargol from neighboring area into the area in front of the fiber tip. One can also see here the initial process of new intrusion formation (outlined with a dashed line). The rate of rise-up front of a given intrusion (which is about 150 μm in average thickness) is found to be described by exponential low (Yusupov et al., 2011b)

$$V = 0.6 \cdot \exp(-1.5 \cdot r), \tag{1}$$

where r is the distance from fiber tip: at 1 mm from fiber tip V = 150 μm/s, while at 2 mm from fiber tip V falls down to 30 μm/s.

The bubbles don't occur up to laser power of 10 W with non-blackened fiber tip and for 0.97 μm laser radiation, while for 1.56 μm laser radiation (which is much stronger absorbed by water) the bubbles are generated at about 1 W of laser power. Blackening of fiber tip results in generation of bubbles for both 0.97 μm and 1.56 μm laser wavelengths.

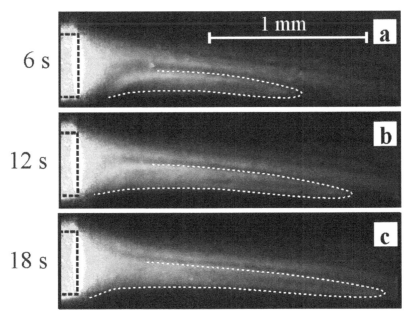

Fig. 4. Microscope pictures (in scattering mode) of intrusions of Ag nanoparticles in water (outlined with dashed line) stimulated by laser induced hydrodynamics nearby optical fiber tip at 1.0 W of 0.97 μm laser power in 6 s (a), 12 s (b), and 18 s (c) of laser irradiation. Fiber tip is shown by dashed line (Yusupov et al., 2011b).

Energy of incident laser light is partly (10–20%) absorbed by the carbon layer on the blackened fiber, so that the fiber is heated. When laser radiation with a power of greater than 3 W is transmitted by the fiber tip in air, the spectrum of the optical radiation from the fiber tip contains the fundamental line (0.97 μm or 1.56 μm) and the broadband visible and near-IR radiation caused by the heating of the tip surface to relatively high temperatures. When a blackened tip is placed into water, the tip surface is effectively cooled and the absence of the broadband radiation means the substantially lower temperatures of the tip surface. However, a medium power laser radiation (1–5 W) is sufficient for surface heating and generation of vapor-gas bubbles. When water is heated, the dissolved gases are liberated in the vicinity of the tip surface and gas bubbles emerge. Water is evaporated inside the bubbles, so that the bubbles are filled with vapor and, consequently, increase in size. At the lower boundary of the above power interval, the bubbles increase in size residing on the tip surface (Fig. 5a). When a critical size is reached, the bubbles are detached and move to the surface.

Water molecules which approach the heated tip surface acquire additional kinetic energy and momentum. The component of the total momentum of vapor molecules that is directed perpendicularly to the tip surface of the fiber towards water appears insufficient for the detachment of the bubble. Figure 5a shows that the bubbles sizes can be close to the diameter of the silica fiber core (400 μm). In the experiments, the bubbles normally emerge at same spots on a tip surface, which correspond to a high temperature areas. Evidently, the presence of such spots is related to the nonuniformity of the carbon layer: the absorbed energy (and, hence, the temperature) is greater for thicker regions. The stabilization (i.e., the

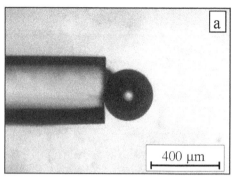
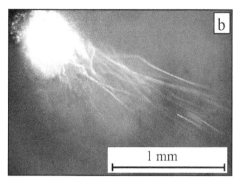

Fig. 5. Laser-induced generation of microbubbles in the vicinity of the blackened end surface of the optical fiber in water for the laser radiation with a wavelength of 0.97 μm and a power of (a) 1 and (b) 5 W. The photograph is taken from above at an exposure time of 250 ms.

attachment of the vapor-gas bubbles to the high temperature spots) can be caused by two reasons. First, the temperature at the hot spot additionally increases owing to the formation of the bubble and the consequent decrease in the local heat sink to water. The second reason is related to the Marangoni effect (Berry et al., 2000): the temperature gradient gives rise to the gradient of surface tension, so that convective flows emerge on the surface of the bubble and cause the force that presses the bubble to the hot spot. The experiments on the growth of the bubbles in the vicinity of the tip surface show that the rate of growth gradually decreases and, finally, the growth is terminated. At a laser power of 1 W, the duration of a relatively fast growth is about 200 ms. Bubble size increases at this stage from zero to 25% of the maximum size. Then, over a few seconds, the growth is well described with the formula (Yusupov et al, 2010):

$$D \propto t^{4/5},$$ (2)

where D is the diameter of bubble and t is time. When laser light is terminated (Fig. 5a), the size of bubble gradually decreases (the bubble remains attached to the tip surface of the fiber) and, finally, the bubble vanishes. Note that a decrease in the size is also non-monotonic. At the first stage with a duration of less than 1 s, the diameter decreases by 8–10%. Then, the slowing takes place. Such a non-monotonic behavior must be related to the fact that the size of bubble decreases at the first stage predominantly, due to a decrease in the temperature of the vapor-gas mixture inside the bubble to the temperature of water in the cell, whereas the second stage is isothermal. The lifetime of such bubbles ranges from 3 to 8 h, and the rate of a decrease in the diameter with time always monotonically increases. At the second stage, the dependence of the diameter on time is well approximated with the formula(Yusupov et al., 2010):

$$D = D_0 \cdot (1 - t / \tau_0)^{\alpha},$$ (3)

where D_0 is the initial diameter, τ_0 is the lifetime, and $a = 0.1$–0.5 is the empirical parameter. Note a similar decrease in the diameter with time at $a = 0.5$ in (Taylor & Hnatovsky, 2004). A qualitatively different scenario corresponds to higher laser powers. The explosive boiling of water is observed in the vicinity of the hot end: the vapor-gas bubbles are ejected from the fiber to water (Fig. 5b) and, then, the velocity decreases due to viscosity. At a finite exposure time the tracks of bubbles moving in water was observed. Notice that the track length corresponds to the mean velocity of the bubble over the exposure time. Bright spots in the vicinity of the tip surface (Fig. 5) are related to stray light: the Vision video camera is sensitive to the near-IR laser radiation.

The side measurements (Fig. 6a) show that the bubbles come to the surface at a certain distance from the fiber. Knowing the vertical velocity of the bubbles (about 5 mm/s in accordance with visual observations) and the trajectories, we can estimate the horizontal velocity (Fig. 6b). The analysis of the trajectories yields an exponential decrease in the horizontal velocity with increasing distance from the fiber: for the slowest and fastest bubbles, we obtain the dependences(Yusupov et al, 2010)

$$V = 67 \cdot \exp(-0{,}82 \cdot r) \tag{4}$$

and

$$V = 101 \cdot \exp(-0{,}74 \cdot r) , \tag{5}$$

respectively, where V is the horizontal velocity in mm/s and r is the distance from the fiber tip surface in millimeters. The relationships show that the velocity of bubbles at the moment of the detachment from the fiber tip ($r = 0$) ranges from 67 to 101 mm/s.

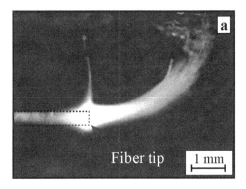 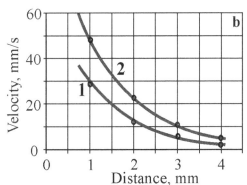

Fig. 6. a - Side view of the tracks of microbubbles in the vicinity of the blackened optical fiber tip surface in water; b - Plots of the horizontal velocity vs. distance from the end surface for slowest (1) and fastest (2) bubbles at a laser wavelength of 0.97 μm and a power of 5 W (Yusupov et al., 2010).

We have directly observed motion of bubbles even in the immediate vicinity of the surface tip (at the maximum velocities) in the experiments on the generation of microbubbles performed with the aid of the Photron Fastcam SA3. Fig. 7 shows the bubbles as dark circles with different sizes. Previous (at time step Δt) positions and sizes are shown as open circles, and the trajectories are shown as rectilinear segments. Table 1 presents the calculated sizes and velocities of the bubbles shown in Fig. 7. It is seen that the bubble with a diameter of 47 μm (bubble 7 in Fig. 7a and Table 1), which is initially located at a distance of about 100 μm from the fiber tip, moves at a mean velocity of 97 mm/s over the observation interval (4.4 ms).

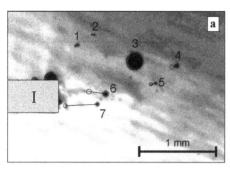 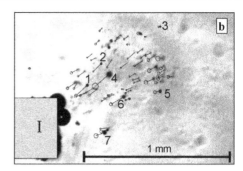

Closed circles 1–7 show positions of bubbles, open circles show previous positions, and rectilinear segments show bubbles trajectories.
The images are taken from above at rates of (a) 10000 and (b) 2000 frames per second.
Laser powers of (a) 3 and (b) 6 W, time intervals Δt = (a) 4.4 and (b) 2.0 ms, and a laser wavelength of 0.97 μm.
The pulse duration is 50 ms and the interval between pulses is 500 ms.

Fig. 7. Displacements of microbubbles (that are generated in the vicinity of the schematically shown blackened tip surface of quartz fiber I) over short time intervals Δt in the presence of laser radiation (Yusupov et al., 2010). a - CW laser radiation. b - Pulsed laser radiation.

Such result is in good agreement with the above estimations of the initial velocities in the vicinity of the fiber tip. The velocities of the bubbles rises rapidly with increasing distance from the fiber: the velocities are not higher than 50 and 20 mm/s at distances of 0.5 mm and 2 mm, respectively (Table 1). When bubbles are generated in a viscous liquid over a relatively long time interval the steady-state flow results in increase of the bubbles velocities. To determine the relative contribution of such a flow, we have measured the motion of microbubbles under the pulsed laser irradiation (Fig. 7b). It is seen that the bubbles predominantly move at relatively large angles relative to the fiber axis. That is caused by the features of the tip surface and hydrodynamic effects. Note that the asymmetry also corresponds to the motion of microbubbles under the continuous wave laser irradiation.

Number of the bubble (Fig. 7)	Parameters of radiation			
	CW radiation, 3 W (Fig. 7a)		pulsed radiation, 6W (Fig. 7b)	
	Diameter, μm	Velocity, mm/s	Diameter, μm	Velocity, mm/s
1	26	9	17	38
2	26	9	10	37
3	200	3	10	5
4	58	16	41	60
5	42	12	21	20
6	63	48	21	52
7	47	97	27	32

Table 1. Parameters of the bubbles shown at Fig. 7

Figure 7b and Table 1 show that a short laser pulse with power of 6 W causes generation of many bubbles, whose diameters range from 10 to 41 µm. The velocities of bubbles are 60 and 20 mm/s in the vicinity of the fiber and at a distance of 300 and 800 µm, respectively. In spite of a twofold increase in the laser power, the maximum velocities of the bubbles in the vicinity of the fiber under the pulsed irradiation are significantly less than the velocities corresponding to the continuous wave irradiation. At a relatively large distance from the fiber end, the velocities corresponding to the pulsed irradiation are also less than the velocities corresponding to the continuous wave irradiation: the velocity of bubble 4 in Fig. 7a is almost equal to the velocity of bubble 5 in Fig. 7b, whose distance from the fiber tip is almost two times shorter. Such result indicates to the presence of water flows in the case of the continuous wave laser irradiation and shows that the flow velocity is comparable with the mean velocity of bubbles.

Such liquid flows are more clearly observed in the microscopic measurements of the laser-induced hydrodynamic effects in the vicinity of the fiber tip surface of the fiber that is placed in a glass capillary filled with water.

3.2 Laser-induced generation of micro-bubbles in a glass capillary

Liquid flows are more clearly observed in the microscopic measurements of the laser-induced hydrodynamic effects in the vicinity of the fiber tip surface of the laser fiber that is placed in the glass capillary filled with water (model of the laser channel).

As it follows from Fig. 8, the attached vapor-gas bubbles at a laser power of 1–2 W emerge at the tip surface and the convective motion is observed in the liquid. A qualitatively different scenario corresponds to a power of 3 W: the microscopic bubbles ejected from the fiber tip move along arc shaped trajectories and entrain liquid flows (Fig. 8a). The intensity of the resulting vortices rapidly increases with increasing radiation power (Fig. 8b). In accordance with the estimations based on the frame-to-frame analysis of the video records, the period of the typical circulating liquid flows at laser powers of 3– 5 W ranges from 0.2 to 1 s. Note that the above effects can be observed in the experiments with the blackened fiber tip at both laser wavelengths (0.97 µm and 1.55 µm). In the absence of the preliminary blackening, the effects are observed only for a radiation wavelength of 1.55 µm. Such a difference is caused by the fact that the radiation with a wavelength of 1.55 µm (unlike the short wavelength

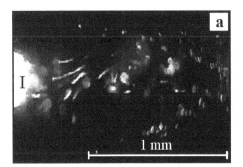 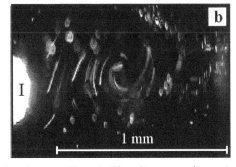

Fig. 8. Water flows that actively circulate inside the glass capillary (with a diameter of 1 mm) in the vicinity of the blackened tip surface I heated by the laser radiation with a wavelength of 0.97 µm and a power of 3 W (a) and 5 W (b)

radiation) is capable of heating a thin water layer in the vicinity of the tip surface to the boiling point, since the absorbance at a wavelength of 1.56 μm is higher than the that at a wavelength of 0.97 μm by a factor of about 20 (Hale & Querry, 1973).

It is possible to visualize the hydrodynamic flows occurring in capillary and caused by laser-induced bubbles generation by microscope visual observing the meniscus. To accomplish this, the silica optical fiber with a 400 μm diameter was introduced into a thin water-filled capillary with a 500 μm internal diameter. The volume of liquid in a capillary was about 20 mm³, and meniscus was located at a 25 mm distance from the fiber tip surface.

Fig. 9 demonstrates the observed variations of a meniscus shape in a glass capillary at a power of laser radiation of 1 W and at laser wavelength of 1.56 μm. Switching of laser radiation on has resulted in growing the distance between optical fiber tip surface, which is caused by the fact that vapor-gas bubbles are formed in a liquid in the course of laser irradiation nearby a fiber tip. Simultaneously with a gradual rise of average volume of liquid in a capillary, quite a strong variations of meniscus shape takes place in this case, which are caused by hydrodynamic processes observing in a capillary water cell. At a certain period of laser irradiation time even water flows occur (Fig. 9b and 9c) caused, presumably, by the appearance and fast motion of quite large vapor-gas bubbles in a water capillary cell. Decrease of laser power causes increase of water streams, and in some cases the eruption of some portion of liquid from a capillary takes place.

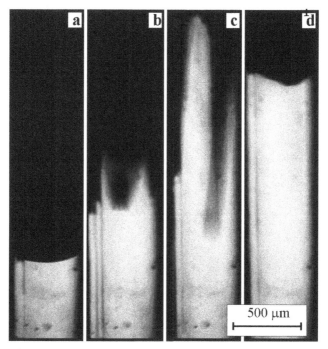

500 μm

Fig. 9. Variation of a meniscus shape in a capillary caused by laser induce hydrodynamics and bubbles formation. Laser wavelength is 1.56 μm, laser power – 1W, internal diameter of capillary - 500 μm.

Knowing the level of the meniscus in a capillary it is possible to determine easily the total volume of vapor-gas bubbles. Fig.10 shows change in the volume of generated bubbles at different laser powers and different laser wavelengths. Our experiments show that the total volume of bubbles rises gradually with time by a logarithmic low after the laser radiation switching on. The total volume at 1 W of laser power rises with time monotonically for both wavelengths, while at higher laser power quite strong fluctuations take place, with the growing in time amplitude. As this takes place, at laser power of 3 W the strong eruption of liquid from the capillary was observed after 4.7 s of laser irradiation (curve 3 at Fig. 10a). At that moment the curve 3 interrupts, since the meniscus went out of visualization zone because of the abrupt decrease of meniscus level.

The total volume of generated bubbles increases with laser power. Comparison of curves 1 and 2 at Fig.10b shows that twofold increase of laser power (from 1 to 2 W) causes about the fourfold rise of the generated bubbles volume. After the laser radiation switching off, the total volume of bubbles first rapidly decreases (vapor condensation inside bubbles), ant next decreases more slowly. It should be noted that quite a strong low-frequency oscillations are observed, caused by variation of total bubbles volume in a capillary.

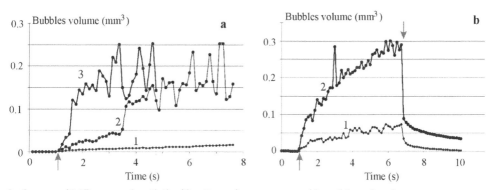

In the case of 0.97 μm wavelength the fiber tip surface was covered by a thin carbon layer.
Arrows show the moments of laser on and laser off.
Digits at curves shows laser power in Watts.

Fig. 10. Change of the total bubbles volume at different powers of lasers with 0.97 μm (a) and 1.56 μm (b) wavelengths of radiation.

Thus, the hydrodynamic processes related to the explosive boiling in the vicinity of the hot tip surface are observed in the liquid even at medium laser powers. Note that the intracapillary liquid exhibits effective mechanical oscillations with a frequency of 1– 5 Hz and appears saturated with microbubbles. We expect the development of such laser-induced hydrodynamic processes in water-saturated biotissues at medium laser powers.

On the one hand, such processes provide the saturation of cavities and fractures in a spinal disc or bone with the water solution containing vapor-gas bubbles. On the other hand, they give rise to high-power acoustic oscillations and vibrations in the organ containing the connective tissue. Apparently, the filling of hernia with vapor-gas bubbles provides the reproducible decrease in the density of herniation immediately after the laser treatment (Sandler et al., 2004; Chudnovskii & Yusupov, 2008).

It is known from (Bagratashvili et al., 2006) that the mechanical action on cartilages in the hertz frequency range actively stimulates the synthesis of collagen and proteoglycans even at relatively small amplitudes. The above estimations show that the pressure on biotissue provided by the vapor-gas bubbles can reach tens of kilopascals. In accordance with (Buschmann et al., 1995; Millward-Sadler & Salter, 2004), such pressures in the hertz frequency range can lead to regenerative processes in cartilage owing to the activation of the interaction of the extracellular matrix with the mechanoreceptors of chondrocytes (integrins).

3.3 Laser-induced generation of bubbles microjets

Note an interesting phenomenon in the experiments on the generation of bubbles in the vicinity of the blackened tip surface of the fiber in the water cell: bubble microjets can be generated at a laser power of less than 3 W (Fig. 11) (Yusupov et al., 2010). The lengths of the microjets (Fig. 11a), which always start in the immediate vicinity of the fiber tip, reach several millimeters, the transverse sizes normally range from 10 to 50 μm, and the sizes of the bubbles that form the jets range from several to ten microns. The lifetime of the microjets ranges from a few fractions of a second to tens of seconds. A microjet that emerges at a certain spot on the tip surface remains attached to this spot and exhibits bending relative to the mean position. Bubble microjets didn't use to be continuous from start to end, the discontinuities used to appear on them, which used to restore quite often. The observations show (Yusupov et al., 2010) that the discontinuities are always related to the hydrodynamic perturbations and are caused by relatively large bubbles that move in the vicinity of the microjet. The appearance of quite a large bubble attached to the fiber tip caused the bubble microjet bending around large bubble (Fig. 11b). Thus, we conclude that two conditions must be satisfied for the generation of the bubble microjets. First, a hot spot must be formed on the tip surface. Second, the neighborhood of such a spot must be free of the centers that provide the generation and detachment of large bubbles. Note that the possibility of bubble microjets in the vicinity of a point heat source is demonstrated in (Taylor & Hnatovsky, 2004).

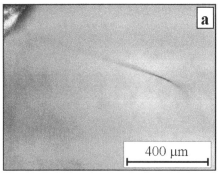
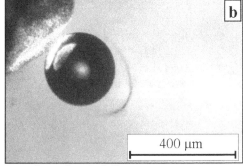

Fig. 11. Bubble microjets in the vicinity of the tip surface of optical fiber.

A part of the blackened fiber tip is sown at the right upper corner.

4. Degradation of optical fiber tip

Laser-induced hydrodynamic effects in water and bio-tissues can lead to the significant degradation of the fiber tip (Yusupov et al., 2011a). The most significant degradation of the fiber tip surface occurs in the regime of channel formation when the fiber is shifted inside the wooden bar that mimics the biotissue. In this case, we observe substantial modifications and distortion of tip surface. The comparison of the sequential photographs (Fig. 12) shows a significant increase in the volume of the fiber fragment (swelling) in the vicinity of fiber tip.

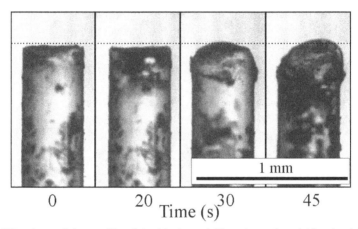

Fig. 12. Modifications of the profile of the blackened fiber tip surface (side view) for regime of channel formation (the channel is formed by the fiber that moves inside the wooden bar with water and the radiation power is 5 W). The left-hand panel shows the original fiber just after its blackening (Yusupov et al., 2011a).

SEM images (Fig. 13) show that the laser action in the regime of the channel formation in the presence of water causes substantial modifications of the working surface: the sharp edge is rounded and surface irregularities (craters) emerge on the tip surface. The image shows that a thin shell (film) with circular holes is formed at the tip surface of the optical fiber. Multiple cracks pass through some of the holes. In addition, we observe elongated crystal-like structures on the surface (Fig. 13b). Looking through the largest hole in the film on the tip surface (at the center of the lower part of the fragment at Fig. 13a), whose dimension in any direction is greater than 10 µm, we observe the inner micron-scale porous structure.

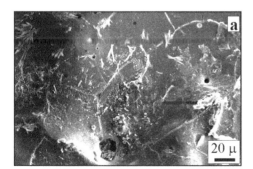

Fig. 13. The microstructure of the fiber tip surface after laser action. a - SEM image of a fragment of the fiber end surface; b - magnified SEM image of a fragment of the end surface with the crystal-like structures on the surface (Yusupov et al., 2011a).

Typical micron-scale circular holes on the film surface (Fig. 13a) can be caused by cavitation collapse of single bubbles. It is well known that cavitation collapse of bubbles in liquid in the vicinity of the solid surface gives rise to the high-speed cumulative microjets which can destroy the solid surface (Suslick, 1994). Apparently, this effect leads to multiple cracks on the film and the formation of the porous structure (Fig. 13a), since the cumulative microjets can punch holes, cause cracks in the film, and destroy the structure of silica fiber tip.

Collapse of cavitation bubble apart from high pressure generation (up to 10^6 MPa) can cause overheating of gas up to temperatures as high as 10^4K. Such high values of water pressure and temperature can result in formation of supercritical water (critical pressure of water is P_c=218 atm, critical temperature - T_c =374°C), which can dissolve silica fiber (Bagratashvili et al., 2009).

Fig. 14 shows Raman spectra of some areas of laser irradiated fiber tip surface (curves 3-5) compared with that of graphite (1) and diamond (2). Raman bands at 1590 cm^{-1} and 1590 cm^{-1} to diamond and graphite nano-phases correspondingly (Yusupov et al., 2011a).

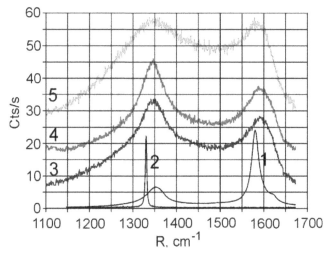

Fig. 14. Raman spectra from different areas of laser fiber tip surface (curves 3, 4 and 5) compared with that of graphite (1) and diamond (2) (Yusupov et al., 2011a).

Formation of diamond nanophase at a fiber tip surface in this case is rationalized by extremely high pressures and temperatures caused by cavitation processes stimulated by laser irradiation (Yusupov et al., 2011a).

5. Laser-induced acoustic effects

Laser-induced hydrodynamics processes in water-saturated bio-tissues causes generation of intense acoustic waves. We have studied the peculiarities of generation of such acoustic waves in water and water-saturated biotissue (intervertebral disc, bone, et al.) in the vicinity of blackened optical fiber tip using acoustic hydrophone (Brul and Kier 8100, Denmark). The hydrophone with 0 – 200 KHz band was placed in water or biotissue at 1cm distance from optical fiber tip. Fig. 15 demonstrates typical example of acoustic response to laser irradiation for two different cases: in the bath of free water (Fig. 15a) and in the case of water

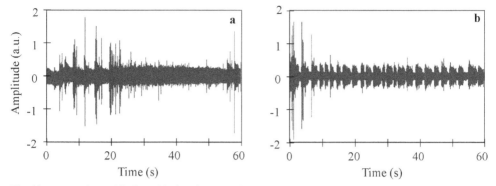

The fiber tip surface is blackened before laser irradiation with 0.97 μm wavelength.

Fig. 15. Fragments of acoustic response to 3 W laser irradiation of water for two different cases: in a bath of free water (**a**) and in a water-filled capillary (**b**).

filled capillary (Fig. 15b). In the case of the bath with free water, the short random laser-induced acoustic spikes take place. At the same time, the acoustic response to laser irradiation in the case of water-filled capillary (which imitates situation in real water-filled biotissue channel) is different (Fig. 15b). Acoustic signal is amplitude-modulated by its feature, and low-frequency modulation period is about 2 s.

Fig. 16 demonstrates acoustic response to laser irradiation of nucleus pulposus *in vivo* when optical fiber was moved forward (regime of channels formation in the course of laser healing of degenerated disc). The acoustic signal is non-stationary by its nature. The short-pulse intense acoustic spikes take place and the signal itself is amplitude modulated (similarly to that in water-filled capillary) with a modulation period of about 3 s.

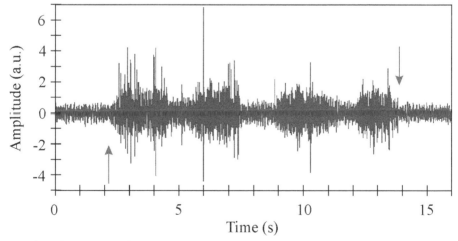

Arrows show the moments of laser on and laser off.

Fig. 16. Acoustic response to 3 W laser irradiation with 0.97 μm wavelength of nucleus pulposus *in vivo,* when optical fiber was moved forward in the intervertebral disc.

The more detailed studies show that for both *in vivo* and *in vitro* cases laser-induced generation of short-pulse intense quasi-periodic acoustic signals. The fragment of spectrogram of acoustic response given at Fig. 17 clearly demonstrates temporal change of spectral components for acoustic signal generated from laser irradiated nucleus pulposus *in vitro* when optical fiber was moved forward in the intervertebral disc (similar to shown at Fig. 1).

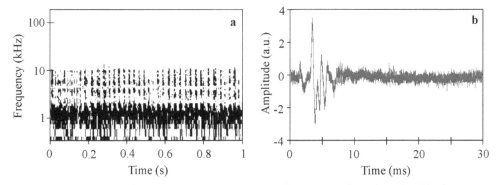

Fig. 17. The fragment of spectrogram (a) ant temporal structure of single pulse (b) of acoustic response generated from laser irradiated nucleus pulposus *in vitro*.

As one can see, the acoustic response in this case has the form of short, intense and broadband (from 0 to 10 kHz) pulses of about 10 ms in duration combined into the series of pulses generated with frequency of 40 Hz. Fig. 17b shows that the amplitude of single pulse is an order of amplitude higher than the background acoustic noise. The most of acoustic power is concentrated in such pulses. The broad spectrum of acoustic pulses and their low duration indicate to shock-type of generated acoustic waves. The acoustic noise has broad spectral maxima in the following spectral intervals: 600 – 700 Hz, 1 - 2 kHz and nearby 10 kHz.

Appearance of these bands are caused by the dynamics of vapor-gas mixture and are associated with acoustic resonances of the system. Notice that laser-induced formation of channels in degenerated spinal discs *in vitro* has been accompanied by 4 Hz in frequency strong visual vibrations of needle with laser fiber.

Generation of such a strong acoustic vibrations is caused in our opinion by contact of overheated (up to >1000 °C (Yusupov et al., 2011a)) fiber tip with water and water-saturated tissue of spinal disc. Such contact can result in explosive boiling of water solution nearby the fiber tip and, also, in burning of collagen in cartilage tissues. Intense hydrodynamic processes can take place nearby optical fiber tip, which are caused by fast heating of water and tissue, by generation and collapse of vapor-gas bubbles (Chudnovskii et al., 2010a, 2010b; Leighton, 1994). As a result, the free space of disc or bone is filled by liquid saturated by vapor-gas bubbles. Resonance vibrations are excited, since both disc and bone are quite good acoustic resonators. These vibrations give rise to low-frequency modulation of acoustic noise (Fig. 16) and to quasi-periodic generation of short intense pulses (Fig. 17) (Chudnovskii et al., 2010a). The acousto- mechanic shock-type processes in resonance conditions results in mixing and transport of gas-saturated degenerated tissue in the space of defect (Chudnovskii et al., 2010b). These processes destroy hernia and decrease its density (Fig. 2b), thus lowering the pressure to nervous roots. Another important impact of such processes is the regeneration of disc tissues through the effects of mechanobiology (Buschmann et al., 1995; Bagratashvili et al., 2006).

6. Formation of filaments

In this division we will show that existence of strongly absorbed agents (in a form of Ag nanoparticles, in particular) in laser irradiated water nearby optical fiber tip can result in appearance of filamentary structures of these agents (Yusupov et al., 2011b). Medium power (0.3 – 8.0 W) 0.97 μm in wavelength laser irradiation of water with added Ag nanoparticles (in the form of Ag-albumin complexes) through 400 μm optical fiber stimulates self-organization of filaments of Ag nanoparticles for a few minutes. These filaments represent themselves long (up to 14 cm) liquid gradient fibers with unexpectedly thin (10 – 80 μm) core diameter. They are stable in the course of laser irradiation, being destroyed after laser radiation off. Such effect of filaments of Ag nanoparticles self-organization is rationalized by the peculiarities of laser-induced hydrodynamic processes developed in water in presence of laser light and by formation of liquid fibers.

Fiber laser radiation (LS-0,97 IRE-Polus, Russia) 0-10 W in output and 0.97 μm in wavelength was delivered into water-filled plastic cell through 400 μm transport silica optical fiber, which was placed horizontally in the cell. Low intensity (up to 1 mW) green pilot beam from the built in diode laser was used to highlight the 0.97 μm laser irradiated zone in the cell. The cell was placed at the sample compartment of optical microscope (MC300, MICROS, Austria) equipped with color digital video-camera (Vision). Spectroscopic studies were performed with fiber-optic spectrum analyzer (USB4000, Ocean Optics) and UV/vis absorption spectrometer (Cary 50, Varian). To measure the refraction index of collargol we have applied the fiber-optic reflectometer FOR-11 (LaserChem, Russia), which provides 10^{-4} precision of refraction index measurements at 1256 nm wavelength. Cleavage of transport optical fiber has been always produced just before each experiment. Ten minutes later (to provide reasonable attenuation of hydrodynamic motions in the cell) the drop (0.01-1 ml in volume) of brown colored collargol (complex of 25 nm in size Ag nanoparticles with albumin) has been smoothly introduced into the water cell 0.5-10 mm aside from the optical fiber tip.

Our in situ optical microscopic studies of laser-induced filament formation were accomplished in two different modes: 1) in transmission mode, using illumination with white light from microscope lamp; 2) in scattering mode, using illumination with green light of pilot laser beam through the same transport fiber.

Experiments show that 0.97 μm fiber laser irradiation of water in the cell with introduced collargol drop causes (in some period of time from seconds to minutes) formation of thin and long quite homogenous filaments, growing along the axis of 0.97 μm laser beam in water. These filaments are brown colored (that gives the evidence of enhanced Ag nanoparticles concentration in filament) and can be seen even with unaided eye.

Fig. 18 demonstrates the microscope image (in transmission mode) of one of such filaments. This filament is located along the axis of output laser beam and is about 17 mm in length. The measured profile of optical density of this filament is triangular in its shape with about the same widths along filament (determined at half-maximum) of ~200 μm.

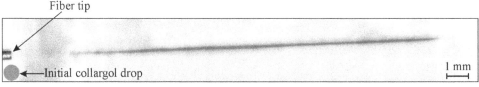

Fig. 18. Micro-image (in transmission mode) of filament of Ag nanoparticles fabricated in water nearby optical fiber tip at 2.5 W of laser power (Yusupov et al, 2011b).

Fig. 19a demonstrates the micro-image of another laser fabricated filament in scattering mode. Intensity of light scattered from this filament decreases gradually with the distance from fiber tip. Attenuation of green light in this case is caused by absorption and scattering of green light in the course of its propagation through the filament. To reveal the peculiarities of filament (given at Fig. 19a) we have performed the following processing of its microscope image: all vertical profiles of image were normalized to local maximum (Fig. 19c); the microscope image was represented in shades of gray (Fig. 19b). As it follows from figures 19b and 19c the length of given filament is about 6 mm, its average width is about 40 μm, and scattering intensity decreases rapidly with the distance from filament axis. Notice that vertical profiles of all fabricated filaments (in both transmission and scattering modes) are almost triangular with a sharp top. It was also established that the end of filament has always a needle-like shape and, also, the width of filament obtained in transmission mode measurements exceeds 3-5 times that obtained in scattering mode.

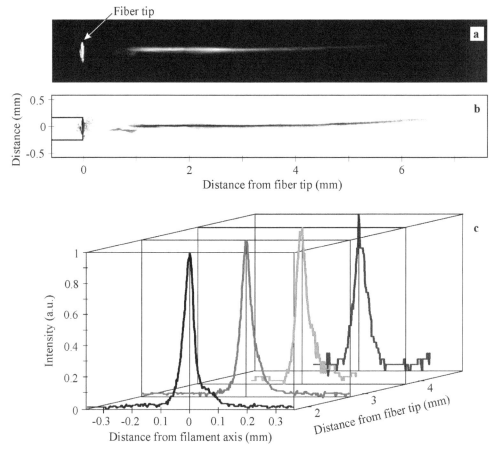

Fig. 19. a - Microscopic picture of filament (in scattering mode) of Ag nanoparticles fabricated in water nearby optical fiber tip at 0.4 W of laser power. b - Image of this filament represented in shades of gray after processing of (see text) of Fig. 19a. c - Normalized vertical profiles of image given at Fig. 19b. (Yusupov et al, 2011b).

It is of importance that filaments of Ag nanoparticles have been formed in our experiments only in the case of existence of initial collargol concentration gradient in laser irradiated water (when collargol drop was introduced initially into water aside from fiber tip). When collargol drop was premixed in water cell before laser irradiation, formation of filaments has never been observed (at any collargol concentrations in the cell and at any laser powers and dozes).

The initial stage of filament self-organization process can be clearly seen in scattering mode (Fig. 4). Some visible hydrodynamic flows take place nearby the fiber tip when laser power is on. Such flows result in intrusion of collargol from neighboring area into the area in front of the fiber tip. The slanting filament structure is clearly seen at Fig. 4. One can also see here the initial process of new intrusion formation (outlined with dashed line). The rate of rise-up front of a given intrusion (which is about 150 μm in average thickness) is found to be described be exponential low (1): at 1 mm from laser fiber tip $V = 1.5 \cdot 10^{-2}$ cm/s, while at 2 mm from laser fiber tip V falls down to $3 \cdot 10^{-3}$cm/s.

We revealed that filaments of Ag nanoparticles self-organized in the course of 0.97 μm laser irradiation can exist in the cell (in the presence of laser beam and with no external mechanical distortions of liquid in the cell) for quite a long period of time. We have supported such filaments for tens of minutes. Notice that both rectilinear and curved filaments were self-organized in our experiments.

After 0.97 μm laser radiation being off, the filaments of Ag nanoparticles have been completely destroyed for 10 – 30 s period of time. Notice that time Δt of diffusion blooming of filament by value, estimated by formula

$$\overline{x}^{-2} = D\Delta t = \frac{kT}{3\pi\mu d}\Delta t \,, \tag{6}$$

where D – is diffusion coefficient of nanoparticle; $k = 1.38 \cdot 10\text{-}23$ J/K – Boltzmann constant; $T(K)$ – absolute temperature; $\mu = 1{,}002 \cdot 10\text{-}3$ (N $\cdot$ s/m²) – dynamic viscosity of water; d=25 nm Ag nanoparticle diameter) gives Δt =25 s for =100 μm.

External mechanical distortions of filament of Ag nanoparticles results in its destruction. However after mechanical distortion being off, the filament can be renewed completely in presence of 0.97 μm laser radiation. Fig. 20 shows the dynamic of such filament renovation after the distortion of self-organized filament (produced by its rapid crossing withthin a metal needle). As one can see from Fig. 20, complete renewal took place for quite a short period of time (~ 20 s).

Our experiments have shown that there is some range of 0.97 μm laser powers for which the effect of laser-induced filament self-organization takes place and is, also, stable and reproducible. At laser powers higher than 8 W we have newer observed filament formation. At 0.2-0.5 W laser power filaments have been formed but have been unstable. The most stable and long-living filaments were observed in 0.5-3 W laser power range. At laser power less than 0.2 W we have never observed such filament formation. The instability of filaments and even their absence at high laser powers is caused by intense laser-induced hydrodynamic processes nearby the fiber tip. Our experiments show that the fiber tip surface is gradually covered by a deposit, which absorbs laser radiation quite well. The wide absorption band of deposit observed at fiber tip can be caused by island film of Ag nanoparticles, and, possibly, by elementary carbon absorption (deposited at fiber tip due to albumin thermo-decomposition). As a result of such deposits, the fiber tip becomes an

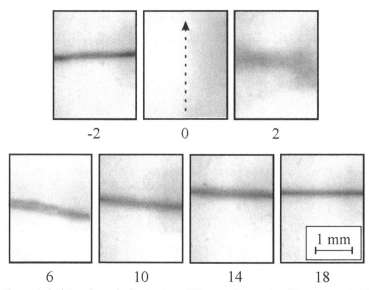

Digits show the period of time from the beginning of filament destruction (Yusupov et al., 2011b).

Fig. 20. Renewal of destroyed filament of Ag nanoparticles in water nearby the tip of optical fiber.

intense heat source. That causes explosive water boiling, intense formation of micro-bubbles, moving rapidly away from fiber tip to liquid (see for example Fig. 1,b) and destroying filament.

We rationalize the observed effect of laser-induced self-organization of filaments from Ag nanoparticles by following mechanisms. Initially (Fig. 21a), laser light absorption by water (the absorption coefficient in water at 0.97 µm is about 0.5 cm-1) causes its heating with the 2-10°C/s rate. Besides, the intense transfer of impulse to water takes place in this case. As a result, the closed axis-symmetric liquid flows are developed being directed from fiber tip. These flows promote Ag nanoparticles intrusion into the laser beam nearby the fiber tip (Fig. 21b). Such intrusions are clearly seen in scattered green laser light (Fig. 4).

Another factor dominates at the second stage of filament self-organization. The refractive index for collargol n_c is higher than that for clean water n_w. The value of n_c-n_w = 0.0044 at wavelength λ=1256 nm was directly measured in our experiments using fiber-optic densitometer. Due to the effect of total internal reflection laser light is concentrated inside intrusion which work in fact as a liquid optical fiber. Channeling of laser light inside intrusion with Ag nanoparticles results in deeper propagation of laser light into water. Light pressure promotes faster movement of intrusion front giving rise to filament (Fig. 21c). As it was shown in (Brasselet et al., 2008), for example, laser light pressure is also able to force through the boundary between two unmixed liquids and to form thin channel of one liquid inside another one, thus forming liquid optical fiber with gradient core. Thus, the image of filament in transmission mode shows optical density of Ag nanoparticles. At the same time the image of filament in scattering mode clearly demonstrate channeling effect in fabricated filament which in fact is a liquid gradient fiber. Such liquid gradient fiber provides also effective channeling of 970nm laser beam, thus promoting filament elongation and spatial stability.

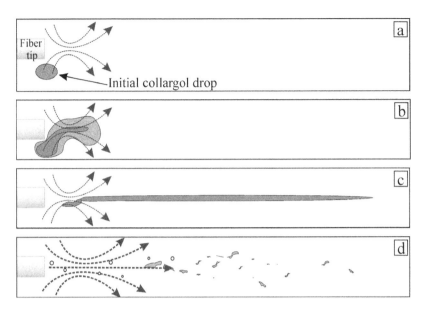

a.   Formation of water flow nearby the fiber tip.
b.   Formation of Ag nanoparticles intrusions.
c.   Fabrication of filaments from Ag nanoparticles.
d.   Intense formation of micro-bubbles, hampering filament formation at high laser power.

Fig. 21. To the explanation of the effect of laser-induced formation of filaments of Ag nanoparticles (Yusupov et al., 2011b).

Laser induced formation of 10-50 µm in thickness and up to few millimeters micro-bubble streams (Fig. 11) can also promote the filaments fabrication observed in our experiments. It is clear, however, that too intense chaotic formation of micro-bubble streams observed at high laser power can hamper filament fabrication (Fig. 21d).
We believe that such filaments of nanoparticles can be developed not only in water media but, also, in other fluids, with other laser wavelength and particles types. The indispensable conditions in this case are the availability of sufficient level of laser light absorption in irradiated medium nearby fiber tip and possibility of liquid fiber formation.

7. Conclusion

Hydrodynamic effects induced by a medium power (1–5 W) laser radiation in the vicinity of the heated fiber tip surface in water and in water-saturated tissues are considered. A threshold character of the dynamics of liquid is demonstrated. At a relatively low laser power (about 1 W), the slow formation of vapor-gas bubbles with sizes of hundreds of microns are observed at the optical fiber tip surface. The bubbles can be attached to the tip surface in the course of laser radiation. At higher laser power increases, effective hydrodynamic processes related to the explosive boiling in the vicinity of the overheated fiber tip surface take place. The resulting bubbles with sizes ranging from a few microns to several tens of microns provide the motion of liquid. The estimated velocities of bubbles in the vicinity of the fiber tip surface can be as high as 100 mm/s. Generation of bubbles in the capillary leads to the circulating water flows with periods ranging from 0.2 to 1 s. Such circulation intensity increases with the laser power. For the laser radiation with a wavelength of 0.97 μm, we observe such effects only for the blackened fiber tip surface, which serves as a local heat source. At a laser power of less than 3 W, stable bubble microjets, which consist of the bubbles (ranging from several to ten microns) can be generated in the vicinity of the blackened tip surface.

Laser-induced hydrodynamic effects in water and bio-tissues can cause the significant degradation of the fiber tip. Cavitation collapse of bubbles in liquid in the vicinity of fiber tip surface gives rise to the high-speed cumulative microjets which can destroy the solid surface. This effect leads to multiple cracks on the film and the formation of the porous structure, formation of supercritical water and even generation of diamonds nano-crystal.

Laser-induced hydrodynamics processes in water and water-saturated bio-tissues are accompanied by generation of intense acoustic waves in resonance conditions, even of shock-type waves. The acousto-mechanic processes results in mixing and transport of gas-saturated degenerated tissue in the space of defect.

We found that medium power (0.3- 8 W) 0.97 μm in wavelength laser irradiation of water with added Ag nanoparticles (in the form of Ag-albumin complexes) through 400μm optical fiber stimulates self-organization of unexpectedly thin (10-80 μm) and lengthy (up to 14 cm) filaments of Ag nanoparticles in the form of liquid gradient fibers. These filaments in water are stable in the course of laser irradiation being destroyed after laser radiation off. Such effect of filaments of Ag nanoparticles self-organization is rationalized by the peculiarities of laser-induced hydrodynamic processes developed in water in presence of laser light.

8. Acknowledgment

This work is supported by Russian Foundation for Basic Research (grant № 09-02-00714).

9. References

Bagratashvili V.N., Sobol E.N., Shekhter A.B. (Eds). (2006). *Laser Engineering of Cartilage*. Fizmatlit, ISBN 5-9221-0729-1, Moscow

Bagratashvili V.N., Konovalov A.N., Novitskiy A.A., Poliakoff M., and Tsypina S.I. (2009). Reflectometric studies of the etching of a silica fiber with a germanium silicate core in sub- and supercritical water. *Russian Journal of Physical Chemistry B, Focus on Physics*, Vol. 3, No. 8, pp. 1154-1164, ISSN 1990-7931

Berry D.W., Heckenberg N.R., and Rubinszteindunlop H. (2000). Effects associated with bubble formation in optical trapping. *Journal of Modern Optics*, Vol. 47, No. 9, pp. 1575 — 1585, ISSN 0950-0340

Brasselet E., Wunenburger R., and Delville J.-P. (2008). Liquid optical fibers with multistable core actuated by light radiation pressure. *Physical Review Letters*, Vol. 101, pp. 1-5, ISSN 1079-7114

Buschmann M.D., Gluzband Y.A., Grodzinsky A.J., and Hunziker E.B. (1995). Mechanical Compression Modulates Matrix Biosynthesis in Chondrocyte Agarose Culture. *Journal of Cell Science*, Vol. 108, pp. 1497-1508, ISSN 0021-9533

Chudnovskii V.M. and Yusupov V.I. (2008). Method of Laser Intervention Effects in Osteochondrosis, Patent RF No. 2321373

Chudnovskii V., Bulanov V., and Yusupov V. (2010a). Laser Induction of Acoustic Hydrodynamical Effects in Medicine. *Photonics*, Vol. 1, pp. 30-36, ISSN 1993-7296

Chudnovskii V.M., Bulanov V.A., Yusupov V.I., Korskov V.I., and Timoshenko V.S. (2010b). Experimental justification of laser puncture treatment of spine osteochondrosis. *Laser Medicine*, Vol. 14, No. 1, pp. 30-35, ISSN 2071-8004

Hale G.M. and Querry M.R. (1973). Optical constants of water in the 200-nm to 200-μm wavelength region. *Applied Optics*, Vol. 12, pp. 555–563, ISSN 0003-6935

Leighton T. G. (Ed.). (1994). *The Acoustic Bubble*, Academic Press Limited, ISBN 0124419208 9780124419209,London

Millward-Sadler S.J. and Salter D.M. (2004). Integrin-dependent signal cascades in chondrocyte mechanotransduction. *Annals of Biomedical Engineering*, Vol. 32, No. 3, pp. 435-446, ISSN 0090-6964

Privalov V.A., Krochek I.V., and Lappa A.V. (2001). Diode laser osteoperforation and its application to osteomyelitis treatment. *Proceedings of the SPIE*, Vol. 4433, pp. 180-185, ISSN 0277-786X

Rokhsar C.K. and Ciocon D.H. (2009). Fractional Photothermolysis for the Treatment of Postinflammatory Hyperpigmentation after Carbon Dioxide Laser Resurfacing. *Dermatologic Surgery*, Vol. 35, No. 3, (March 2009), pp. 535-537, ISSN 1524-4725

Sandler B.I., Sulyandziga L.N., Chudnovskii V.M., Yusupov V.I., and Galin Y.M. (2002). *Bulletin physiology and pathology of respiration*, Vol. 11, pp. 46-49, ISSN 1998-5029

Sandler B.I., Sulyandziga L.N., Chudnovskii V.M., Yusupov V.I., Kosareva O.V., and. Timoshenko V.C. (2004). *Prospects for Treatment of Compression Forms of Discogenic Lumbosacral Radiculitis by Means of Puncture Nonendoscopic Laser Operations* (Skoromec A.A.), Dalnauka, ISBN 5-8044-0443-1, Vladivostok

Suslick K.S. (1994). The chemistry of ultrasound. *The Yearbook of Science & the Future*, pp 138-155, Encyclopaedia Britannica, ISBN 0852294026, Chicago

Taylor R.S. and Hnatovsky C. (2004). Growth and decay dynamics of a stable microbubble produced at the end of a near-field scanning optical microscopy fiber probe. *Journal of Applied Physics*, Vol. 95, No. 12, (June 2004), pp. 8444-8449, ISSN 0021-8979

Van den Bos R., Arends L., Kockaert M., Neumann M., Nijsten T. (2009). Endovenous therapies of lower extremity varicosities: a meta-analysis. *Journal of Vascular Surgery*, Vol. 49, No. 1, pp. 230-239, ISSN 0741-5214

Yusupov V.I., Chudnovskii V.M., and Bagratashvili V.N. (2010). Laser-induced hydrodynamics in water-saturated biotissues. 1. Generation of bubbles in liquid. *Laser Physics*, Vol. 20, No. 7, pp.1641-1646, ISSN 1054 660X

Yusupov V.I., Chudnovskii V.M., and Bagratashvili V.N. (2011a). Laser-induced hydrodynamics in water-saturated biotissues. 2. Effect on Delivery Fiber. *Laser Physics*, Vol. 21, No. 7, pp. 1230-1234, ISSN 1054 660X

Yusupov V.I., Chudnovskii V.M., Kortunov I.V., Bagratashvili V.N. (2011b). Laser-induced self-organization of filaments from Ag nanoparticles. *Laser Physics Letters*, Vol. 8, No. 3, (March 2011), pp. 214–218, ISSN 1612-2011

6

Endocrine Delivery System of NK4, an HGF-Antagonist and Anti-Angiogenic Regulator, for Inhibitions of Tumor Growth, Invasion and Metastasis

Shinya Mizuno[1] and Toshikazu Nakamura[2]
[1]Division of Virology, Department of Microbiology and Immunology,
Osaka University Graduate School of Medicine, Osaka
[2]Kringle Pharma Joint Research Division for Regenerative Drug Discovery,
Center for Advanced Science and Innovation, Osaka University, Osaka
Japan

1. Introduction

Estimates of the worldwide incidence and mortality from 27 cancers in 2008 have been prepared for 182 countries by the International Agency for Research on Cancer (Ferlay *et al.*, 2010). Overall, an estimated 12.7 million new cancer cases and 7.6 million cancer deaths occur in 2008, with 56% of new cancer cases and 63% of the cancer deaths occurring in the less developed regions of the world. The most commonly diagnosed cancers worldwide are lung (1.61 million, 12.7% of the total), breast (1.38 million, 10.9%) and colorectal cancers (1.23 million, 9.7%). Cancer is neither rare anywhere in the world, nor mainly confined to high-resource countries. Many cancer subjects die from cancer as a result of organ failure due to "metastasis" (Geiger & Peeper, 2009), thus indicating that medical control of tumor metastasis leads to a marked improvement in cancer prognosis.

The acquisition of the metastatic phenotype is not simply the result of oncogene mutations, but instead is achieved through an interstitial stepwise selection process (Mueller & Fusenig, 2004). The dissociation and migration of cancer cells, together with a breakdown of basement membranes between the parenchyme and stroma, are a prerequisite for tumor invasion. The next sequential events involved in cancer metastasis include the following: (i) penetration of cancer cells to adjacent vessels (*i.e.*, intravasation); (ii) suppressed anoikis (*i.e.*, suspension-induced apoptosis) of cancer cells in blood flow; and (iii) an extravascular migration and re-growth of metastatic cells in the secondary organ. For an establishment of anti-metastasis therapy, it is important to elucidate the basic mechanism(s) whereby tumor metastasis is achieved through a molecular event(s).

Hepatocyte growth factor (HGF) was discovered and cloned as a potent mitogen of rat hepatocytes in a primary culture system (Nakamura *et al.*, 1984, 1989; Nakamura, 1991). Beyond its name, HGF is now recognized as an essential organotrophic regulator in almost all tissues (Nakamura, 1991; Rubin *et al.*, 1993; Zarnegar & Michalopoulos, 1995; Birchmeier & Gherardi, 1998; Nakamura & Mizuno, 2010). Actually, HGF induces mitogenic, motogenic and morphogenic activities in various types of cells via its receptor, MET (Bottaro *et al.*, 1991; Higuchi *et al.*, 1992). HGF is required for organogenesis in an embryonic stage and for tissue repair in adulthood during various diseases (Nakamura, 1991; Birchmeier & Gherardi, 1998; Nakamura & Mizuno, 2010). Several lines of *in vitro* studies indicate that HGF stimulates scattering and migration of cancer cells (Matsumoto *et al.*, 1994, 1996a; Nakamura *et al.*, 1997). In malignant tumors, HGF is expressed by stromal cells, such as fibroblasts, while MET is over-expressed by cancer cells, thus suggesting in the mid-1990s that a paracrine signal from HGF-producing stroma cells to carcinomas may cause malignant behaviors, such as invasion and metastasis (Matsumoto *et al.*, 1996b).

NK4 is an intra-molecular fragment of HGF, which is generated by a chemical cleavage of mature form HGF (Date *et al.*, 1997; Nakamura *et al.*, 2010). NK4 includes an N-terminal hairpin domain and 4-kringle domains (K1-K4) of HGF α-chain, which binds to MET. Thus, NK4 antagonizes HGF activities as a competitive inhibitor. Using NK4 as an HGF-antagonist in rodents with malignant tumors, we have accumulated evidence showing that endogenous HGF-MET cascade is a key conductor for tumor metastasis, while inhibition of MET signals leads to the arrests of tumor growth. Unexpectedly, NK4 prohibits tumor angiogenesis through a MET-independent mechanism. This review focuses on the roles of HGF in cancer biology and pathology. We also emphasize the effectiveness of NK4 in experimental cancer models where NK4 is supplemented via a "hydrodynamics-based" gene therapy.

2. Effects of HGF on intra-tumor cells during cancer progression

In the mid-1980s, MET was identified as a mutated oncogene from carcinogen-induced osteosarcoma cells (MNNG-HOS) that transform NIH3T3 fibroblasts (Cooper *et al.*, 1984). MET-encoding protein has a tyrosine kinase activity (Dean *et al.*, 1985), suggesting that MET may be an orphan receptor of growth factors. In the early 1990s, MET-coding product was demonstrated to be a high-affinity receptor for HGF (Bottaro *et al.*, 1991; Higuchi *et al.*, 1992). Scatter factor (SF) stimulates tumor cell movement, as its name indicates, and is shown molecularly identical to HGF (Konishi *et al.*, 1991; Weidner *et al.*, 1991). HGF has several activities required for tumor cell invasion and metastasis, as described below. In this section, we summarize the direct effects of HGF on intra-tumor cells, including carcinoma, and on vascular and lymphatic cells prior to discussion of the contribution of HGF-MET cascades during tumor malignancy.

2.1 Scattering and migration of tumor cells

Initial events for the metastatic spread of tumors involve loss of cell-cell contact within the primary tumor mass. The integrity and morphology of epithelial tumor cell colonies are maintained by cell-cell contact mediated by cadherins and its associated intracellular catenin molecules. Cancer cells must lose their tight cell-to-cell contact by down-regulation of cadherin-cadherin complex during invasion into adjacent tissues. HGF induces scattering (*i.e.*, dispersion of cluster cells into single cells) via an endocytosis of E-cadherin from cell surface to cytoplasma (Watabe *et al.*, 1993; Miura *et al.*, 2001). During cell migration, HGF activates the Ras-Rab5 pathway for endocytosis of cadherins (Kimura *et al.*, 2006), which triggers nuclear localization of β-catenin, a transcription factor of genes responsible for cell motility (Hiscox & Jiang, 1999). Stimulation of an Rho small G protein cascade and activation of cdc42, rac and PAK by HGF leads to the disassembly of stress fiber or focal adhesions, while lamellipodia

Endocrine Delivery System of NK4, an HGF-Antagonist and Anti-Angiogenic Regulator, for Inhibitions of Tumor
Growth, Invasion and Metastasis

121

formation and cell spreading are enhanced by HGF (Royal *et al.*, 2000). These changes confer a down-stream mechanism of MET-mediated cancer invasion.

2.2 Breakdown of basement membranes

During cancer invasion, tumor cells must move across a basement membrane between epithelium and lamina propria (*i.e.*, sub-epithelium). HGF stimulates motility in a biphasic process: cells spread rapidly and form focal adhesions, and then they disassemble these condensations, followed by increased cell locomotion. In the early phase (*i.e.*, within a few minutes post-stimulation), HGF induces phosphorylation of focal adhesion kinase (FAK) together with a tight bridge between the extra-cellular matrix (ECM) and integrins of cancer cells (Matsumoto *et al.*, 1994; Parr *et al.*, 2001). In the later phase, HGF-stimulated cancer cells invade into matrix-based gels *in vitro*, or across basement membrane ECM *in vivo* (Nakamura *et al.*, 1997). In this process, HGF up-regulates several types of matrix metalloproteinase (MMP), such as MMP-1, -2, and -9, through activation of Ets, a transcriptional factor of MMPs (Li *et al.*, 1998; Nagakawa *et al.*, 2000; Jiang *et al.*, 2001). Considering that MMP-inhibitors diminish HGF-mediated migration, the induction of MMP through HGF-Ets cascade is essential for tumor invasion into adjacent normal tissues.

2.3 Endothelial attachment and extravasation of cancer cells

Needless to say, tumor angiogenesis as well as lymphatic vessel formation are important for delivery of cancer cells from the primary tumor to secondary organs. HGF enhances angiogenesis via induction of the proliferation and morphogenesis of endothelial cells (EC) (Bussolino *et al.*, 1992; Nakamura *et al.*, 1996). Actually, HGF supplementation leads to the enhancement of tumor angiogenesis *in vivo* (Laterra *et al.*, 1997). Recent studies delineated the capacity of HGF to induce lymphatic morphogenesis (Kajiya *et al.*, 2005; Saito *et al.*, 2006). Thus, HGF is considered to facilitate cancer metastasis via neo-induction of vascular or lymphatic vessel beds. HGF has a direct effect on EC for enhancing tight adhesion of tumor cells on endothelium via FAK phosphorylation (Kubota *et al.*, 2009a). Furthermore, HGF decreases endothelial occludin, a cell-cell adhesion molecule (Jiang *et al.*, 1999a). Under such a loss of EC-EC integrity, HGF decreases the trans-endothelial resistance of tumor vessels and enhances cancer invasion across an EC barrier (*i.e.*, intravasation in primary tumors and extravasation in metastatic organs) (**Fig. 1**).

2.4 Prevention of cancer cell anoikis

Anoikis, also known as suspension-induced apoptosis, is a term used to describe programmed cell death (apoptosis) of epithelial cells induced by loss of matrix attachment. In addition to gaining functions of invasion and angiogenesis, cell resistance to anoikis also appears to play an important role in tumor progression and metastasis as tumor cells lose matrix attachment during metastasis. However, it is unknown how cancer cells escape from anoikis-like death during metastasis. It was demonstrated, in a non-adherent culture models, that HGF is a key molecule inhibiting suspension-induced anoikis, and this effect is mediated via a crosstalk that is, in turn, mediated by phosphatidyl-inositol 3-kinase (PI-3K) and extracellular signal-regulated kinase (ERK)-1/2 (Zeng *et al.*, 2002; Kanayama *et al.*, 2008). A recent report described that tetraspanin CD151-knockdown abolishes preventive effect of HGF on tumor anoikis (Franco *et al.*, 2010). Thus, it is likely that cell surface tetraspanins are important for signaling complexes between MET and integrin-β4, a known amplifier of HGF-mediated cell survival.

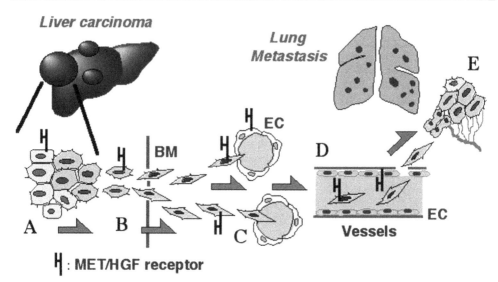

Fig. 1. Various effect of HGF on cancer cells and endothelial cells (EC) during tumor progression. For example, sequential events during the lung metastasis of hepatic carcinoma are summarized as follows: (A) dissociation and scattering of hepatocellular cancer cells through an HGF-induced endocytosis of cadherins; (B) tumor migration into stromal areas across the basement membrane (BM) is mediated via MMP-dependent matrix degradation and Rho-dependent cell movement; (C) invasion of tumor cells into neighboring vessels (*i.e.*, intravasation) where the tight junction between ECs is lost by HGF-MET signaling; (D) inhibition of tumor cell anoikisis by MET-AKT cascades during blood flow, and out-flux of tumor cells across vessel walls (*i.e.*, extravasation); and (E) in the lung, HGF supports growth of metastatic nodules via providing vascular beds as an angiogenic factor.

Overall, HGF is shown to take direct action on carcinoma cells: (i) cell spreading via an endocytosis of cadherins; (ii) enhancement of invasion across basement membranes via Rho-dependent and MMP-dependent pathways; and (iii) anti-anoikis activity during blood circulation. Toward tumor vessels, HGF elicits vascular and lymphatic EC proliferation and branching angiogenesis, while intravasation and extravasation are achieved through HGF-induced reduction of EC-EC integrity. These HGF-MET-mediated biological functions seem advantageous for invasion and metastasis of malignant tumors, including carcinoma and sarcoma (**Fig. 1**).

[Note] Long-term administration of recombinant HGF does not elicit tumor formation in healthy animals, and this result supports a rationale of HGF supplement therapy for treating chronic organ diseases, such as liver cirrhosis, at least in cancer-free patients.

3. Regulation of HGF production by cancer cells

Several lines of histological evidence indicate that HGF is produced in stroma cells, such as fibroblasts, vascular EC and smooth muscle cells in tumor tissues. In contrast, MET is over-expressed mainly by tumor cells, particular near invasive areas, implying a possible paracrine signal from HGF-producing stroma cells to MET-expressing carcinoma cells

(Matsumoto *et al.*, 1996b). Herein, we will discuss the molecular basis whereby stromal HGF production is up-regulated by tumor cells during cancer invasion and metastasis.

3.1 Stroma as a microenvironment to determine behaviors of tumors

The important roles of stroma during tumor progression are demonstrated through several independent studies. Carcinoma-associated fibroblasts, but not normal fibroblasts, stimulate tumor progression of initiated non-tumorigenic epithelial cells both in an *in vivo* tissue recombination and in an *in vitro* co-culture system (Olumi *et al.*, 1999). Transforming growth factor (TGF)-β signaling is critical for down-regulating HGF production (Matsumoto *et al.*, 1992). Of note, an inactivation of TGF-β type II receptor gene in stromal fibroblasts leads to the onset of epithelial growth and invasion (Bhowmick *et al.*, 2004). In this process, activation of paracrine HGF is a key mechanism for stimulation of epithelial proliferation (Bhowmick *et al.*, 2004). Thus, the suppression of HGF production by TGF-β seems to be important for an escape from cancer metastasis (Matsumoto & Nakamura, 2006).

3.2 Regulation of HGF production in stroma by tumor cells

As repeated, a major source of HGF in tumors is stromal cells (including fibroblasts, endothelium, macrophages and neutrophils) (Wislez *et al.*, 2003; Matsumoto & Nakamura, 2006; Grugan *et al.*, 2010). Thus, how stromal HGF is up-regulated during tumor progression should be discussed. There is now ample evidence that numerous types of carcinoma cells secrete soluble factors that induce HGF production in stromal cells (*i.e.*, HGF-inducers). For example, conditioned medium obtained from breast cancer cells enhances HGF production in fibroblasts, along with a raise in prostaglandin-E2 (Matsumoto-Taniura *et al.*, 1999). Of note, suppression of prostaglandin-E2 production by indomethacin leads to down-regulation of stromal HGF production and suppression of tumor migration *in vitro* (Matsumoto-Taniura *et al.*, 1999), indicating that cancer-derived prostaglandins are important for up-regulating HGF in stromal cells (Matsumoto-Taniura *et al.*, 1999; Pai *et al.*, 2003). Other carcinoma-derived HGF-inducers are interleukin-1β (IL-1β), basic fibroblast growth factor (b-FGF), platelet-derived growth factor (PDGF), and TGF-α (Hasina *et al.*, 1999; Matsumoto & Nakamura, 2003). These results indicate a crosstalk between carcinoma and stroma, mediated via a paracrine loop of HGF-inducers produced by carcinoma and HGF secreted from stroma cells, such as fibroblasts (Matsumoto *et al.*, 1996a).

3.3 Inflammation-mediated HGF up-regulation mechanism

In addition to stromal fibroblasts, tumor-associated macrophages (TAM) are known to highly produce HGF during non-small lung cancer invasion (Wang *et al.*, 2011). It is reported that TAM isolated from 98 primary lung cancer tissues show the higher production of HGF, along with the concomitant increases in urokinase-type plasmin activator (uPA), cyclooxygenase-2 (Cox2) and MMP-9 (Wang *et al.*, 2011). Anti-MMP-9 antibody largely diminishes TAM-induced invasion, while Cox2 and uPA are critical for HGF production and activation, respectively, suggesting that Cox2-uPA-HGF-MMP cascades in TAM participate in non-small lung cancer invasion. Likewise, HGF production is enhanced by neutrophils infiltrating bronchiolo-alveolar subtype pulmonary adenocarcinoma (Wislez *et al.*, 2003).

Clinical studies demonstrate that serum levels of HGF are elevated in patients with recurrent malignant tumors (Wu *et al.*, 1998; Osada *et al.*, 2008), thus suggesting an endocrine mechanism of the HGF delivery system. In this regard, it is known that peripheral blood monocytes produce HGF, contributing to the increase in blood HGF levels via an endocrine mechanism (Beppu *et al.*, 2001). Overall, production of HGF by inflammatory cells is involved in carcinoma invasion and metastasis (*i.e.*, local system), while peripheral blood monocytes seem to prevent tumor cell anoikis during metastasis, possibly by a release of HGF into blood (*i.e.*, systemic system).

4. Structure and activity of NK4 as HGF antagonist

HGF is a stromal-derived paracrine factor that has stimulated cancer invasion at least *in vitro* (Matsumoto *et al.*, 1994; Matsumoto *et al.*, 1996a; Nakamura *et al.*, 1997). Clinical studies suggest that the degree of serum HGF and Met expressions in cancer tissues appears to correlate with a given prognosis (Yoshinaga *et al.*, 1993; Osada *et al.*, 2008). Thus, it is hypothesized that *in vivo* inhibition of HGF-MET signaling may be a reasonable strategy to prohibit cancer metastasis. To test this hypothesis, we prepared NK4 as an intra-molecular fragment of HGF via a chemical digestive process (Date *et al.*, 1997; Matsumoto *et al.*, 1998). As expected, NK4 bounded to MET and inhibited HGF-MET coupling as a competitive inhibitor. An additional "unexpected" value was that NK4 inhibited tumor angiogenesis via a MET-independent pathway. This section focuses on the biological value of NK4 as an HGF-antagonist and as an angiogenesis inhibitor.

4.1 Structure and anti-invasive function of NK4

NK4 was initially purified as a fragment from elastase-digested samples of recombinant human HGF (Date *et al.*, 1997). The N-terminal amino acid sequence of NK4 and of the remnant fragment, assumed to be composed of an HGF β-chain, revealed that NK4 is cleaved between the 478th valine and the 479th asparagine. The N-terminal amino acid sequence of NK4 revealed that the N-terminal structure of NK4 is the same as undigested HGF (*i.e.*, 32nd pyroglutamate), indicating that NK4 is composed of the N-terminal 447 amino acids of the α-chain of HGF and contains the N-terminal hairpin domain and four kringle domains (thus designated NK4) (**Fig. 2A**). The binding domains that are responsible for high-affinity binding to MET are the N-terminal hairpin and the first kringle domains in NK4 (and HGF). MET tyrosine phosphorylation occurs in A549 lung carcinoma within 10 minutes after HGF addition, while NK4 inhibits the HGF-mediated MET activation (**Fig. 2B**). Actually, NK4 functions as an HGF-antagonist: HGF induces invasion and migration of the gallbladder and bile duct carcinoma cells in ECM-based gels, while NK4 inhibits HGF-induced invasion in a dose-dependent manner (**Fig. 2C**) (Date *et al.*, 1998). These anti-invasive effects of NK4 are seen in distinct types of cancer cells (Hiscox *et al.*, 2000; Maehara *et al.*, 2001; Parr *et al.*, 2001), strengthening the common role of NK4 during cancer migration.

4.2 Perlecan-dependent anti-angiogenic mechanism by NK4

Vascular EC highly express MET, while HGF stimulates mitogenic and morphogenic activities in EC (Nakamura *et al.*, 1996), thus suggesting that NK4 could inhibit HGF-induced angiogenesis. Actually, NK4 potently inhibited the HGF-mediated proliferation of EC *in vitro* (Jiang *et al.*, 1999b). Strikingly, NK4 also inhibited microvascular EC proliferation and migration, induced by other angiogenic factors, such as b-FGF and vascular endothelial

Endocrine Delivery System of NK4, an HGF-Antagonist and Anti-Angiogenic Regulator, for Inhibitions of Tumor Growth, Invasion and Metastasis

125

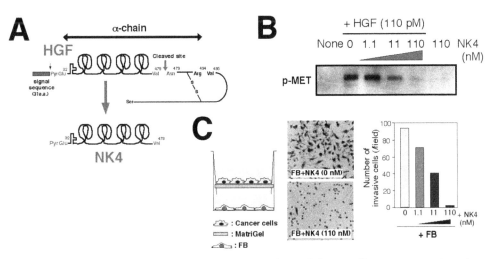

Fig. 2. Preparation of NK4 as an HGF-antagonist and its inhibitory effects on tumor invasion *in vitro*. (A) Preparation and structure of NK4. NK4 is generated via a cleavage of HGF between 478th Val and 479th Asn. (B) Inhibition of HGF-mediated MET tyrosine phosphorylation by NK4 in lung carcinoma cells. (C) Biological activity of NK4. Cancer cell invasion (upper chamber) is induced across a Matrigel layer when fibroblasts (FB) are placed on a lower chamber. In this co-culture system, NK4 inhibits FB-induced tumor cell invasion in a dose-dependent manner.

growth factor (VEGF) (**Fig. 3A**) (Kuba *et al.*, 2000). When a pellet containing b-FGF was implanted under the rabbit cornea, angiogenesis was rapidly induced. In this model, NK4 inhibited b-FGF-induced angiogenesis (**Fig. 3B**). *In vitro* models of EC proliferation, HGF and VEGF phosphorylate MET and KDR/VEGF receptor, respectively, whereas NK4 inhibits HGF-induced MET tyrosine phosphorylation, but not VEGF-induced KDR phosphorylation (Kuba *et al.*, 2000). Nevertheless, NK4 inhibited the VEGF-mediated EC proliferation without modification of VEGF-mediated ERK1/2 (p44/42 mitogen-activated protein kinase) activation. These results suggest the presence of another mechanism whereby NK4 inhibits VEGF- and b-FGF-mediated angiogenesis.

The fibronectin-integrin signal is essential for the spreading and proliferation of EC. Based on this background, we demonstrated that NK4 mediated growth arrest of EC is due to a loss of the fibronectin-integrin signal. Affinity purification with NK4-immobilized beads revealed that NK4 binds to perlecan (Sakai *et al.*, 2009). Consistent with this result, NK4 was co-localized with perlecan in EC. Perlecan is a multi-domain heparan sulfate proteoglycan that interacts with basement membrane components such as fibronectin. Of interest, knockdown of perlecan expression by siRNA diminished the fibronectin assembly and EC spreading, indicating an essential role of fibronectin-perlecan interaction during EC movement. A recent report described that NK4-perlecan interaction suppressed the normal assembly of fibronectin by perlecan (Sakai *et al.*, 2009). As a result, FAK activation became faint in EC after NK4 treatment. Under such a loss of fibronectin-integrin signaling by NK4, EC growth and motility were suppressed, even in the presence of b-FGF or VEGF. This is the reason why NK4 arrests b-FGF- or VEGF-mediated angiogenesis (**Fig. 3C**).

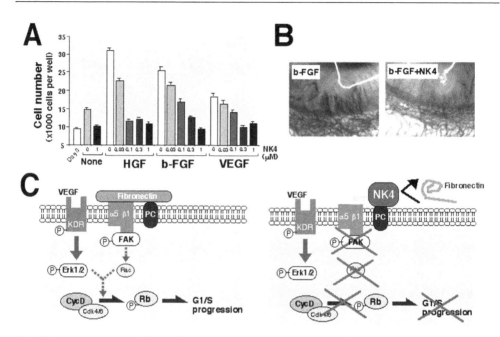

Fig. 3. Anti-angiogenic effects of NK4 via a perlecan-dependent mechanism. (A) NK4 suppresses HGF-, b-FGF-, and VEGF-induced proliferation of EC *in vitro* (Kuba *et al.*, 2000). (B) Inhibition of b-FGF-induced corneal neovascularization by NK4 treatment in rabbits. (C) Involvement of perlecan (PC) in NK4-mediated growth arrest of EC. Left: Cell surface PC is required for the binding of fibronectin and α5β1-integrin, leading to FAK phosphorylation and crosstalk of VEGF-VEGF receptor (KDR) signaling. Right: NK4 binds to PC, and then the binding of integrin to fibronectin is impaired. As a result, VEGF fails to elicit G1/S progression of EC in the presence of NK4 (Sakai *et al.*, 2009).

We have accumulated *in vitro* evidence showing that HGF-MET system may elicit cancer invasion via a paracrine loop of stroma-carcinoma interaction. This phenomenon is also demonstrated *in vivo*: anti-HGF antibody potently suppressed the tumor invasion in a mouse model of pancreas cancer (Tomiola *et al.*, 2001). On the other hand, several investigators proposed, in the late-1990's, a new concept that tumor angiogenesis inhibition leads to the arrest of cancer growth and metastasis (Yancopoulos *et al.*, 1998). Inhibition of tumor angiogenesis leads to local hypoxia, and then apoptotic death of cancer cells is associated with the arrests of tumor growth and metastasis (*i.e.*, cytostatic therapy). In this regard, NK4 also elicits an anti-angiogenic effect via perlecan-dependent mechanism. Thus, bi-functional properties of NK4 as an HGF antagonist and angiogenesis inhibitor raise a possibility that NK4 may prove therapeutic for cancer patients, as follows.

5. Anti-cancer therapy using NK4 in animal models

Carcinoma and sarcoma show malignant phenotypes prompted by a stroma-derived HGF-MET signal at least *in vitro*. If NK4 could block MET signaling as an HGF-antagonist *in vivo*, supplemental therapy with NK4 would be a pathogenesis-based strategy to counteract tumor invasion and metastasis. This hypothesis is widely demonstrated through extensive studies using tumor-bearing animals, as described below.

5.1 First evidence of NK4 for inhibition of carcinoma progression *in vivo*

HGF, or co-cultured fibroblasts, are known to induce invasion of gallbladder carcinoma cells (GB-b1) across Matri-gel basement membrane components (Li *et al.*, 1998). NK4 competitively inhibits the binding of HGF to MET on GB-d1 cells. As a result, NK4 diminishes HGF-induced, or fibroblast-induced, motogenic activities (Date *et al.*, 1998), thus suggesting that stroma-derived HGF is a key conductor for provoking tumor invasion. Such an important role of HGF was also demonstrated *in vivo*. Subcutaneous inoculations of human gallbladder carcinoma GB-d1 cells in nude mice allow for primary tumor growth and invasion to adjacent muscular tissues. Using this conceptual model, we provided the first evidence of NK4 as an anti-tumor drug (Date *et al.*, 1998). Recombinant NK4 has inhibited the growth and muscular invasion in a mouse model of gallbladder carcinoma. Consistent with tumor growth arrest, apoptotic change becomes evident during NK4 injections. Since HGF has an anti-apoptotic effect on cancer cells (Zeng *et al.*, 2002), reverse of HGF-induced protection by NK4 may be one of the mechanisms whereby carcinoma growth can be suppressed during NK4 supplemental therapy.

5.2 Inhibition of tumor angiogenesis by NK4 treatment

In a culture of EC, NK4 produces anti-angiogenetic effects via a MET-independent pathway (Kuba *et al.*, 2000; Nakabayashi *et al.*, 2003). These effects are also observed in animal models of malignant tumors: administration of recombinant NK4 suppressed primary tumor growth, metastasis of Lewis lung carcinoma, and Jyg-MC(A) mammary carcinoma implanted into mice (Kuba *et al.*, 2000), although neither HGF nor NK4 affected proliferation and survival of these tumor cells *in vitro*. NK4 treatment resulted in a remarkable decrease in microvessel density and an increase in apoptotic tumor cells in primary tumors, suggesting that the inhibition of tumor growth by NK4 may be achieved by the suppression of tumor angiogenesis (Kuba *et al.*, 2000). The anti-angiogenic effects of NK4 are widely demonstrated in various types of cancers [see our review articles (Matsumoto & Nakamura, 2005; Matsumoto *et al.*, 2008a,b)]. Because the inhibition of angiogenesis by NK4 leads to tumor hypoxia, hypoxia-primed apoptosis may contribute to a reduction in tumor size during NK4 supplemental therapy.

5.3 Delayed NK4 therapy for attenuation of end-stage pancreas carcinoma

Anti-tumor effect of NK4 is also observed in a mouse model of advanced pancreas carcinoma (Tomioka *et al.*, 2001). When NK4 treatment was initiated on day 10, a time when cancer cells were already invading surrounding tissues, NK4 potently inhibited the tumor growth, peritoneal dissemination, and ascites accumulation at 4 weeks after the inoculation. Such an anti-tumor effects of NK4 correlated with decreased vessel density in pancreatic tumors. In an end-stage of pancreas cancer, NK4 inhibited the malignant phenotypes, such as peritoneal dissemination, invasion of cancer cells into the peritoneal walls and ascites accumulation (Tomioka *et al.*, 2001). As a result, NK4 prolonged the survival time of mice at an end-stage of cancer (**Fig. 4**). Because effective systemic therapy for pancreatic cancer is currently not available, and diagnosing pancreatic cancer in its early stages is difficult, the highly invasive and metastatic behaviors of pancreatic cancer lead to difficulty in attaining a

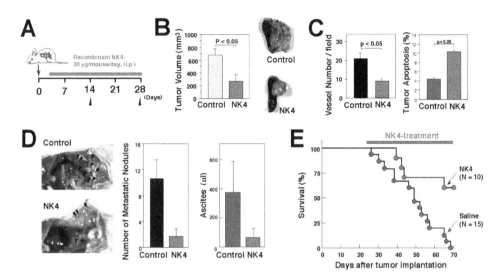

Fig. 4. Anti-tumor effects of NK4 on advanced pancreas cancer in mice. (A) Schedules for NK4 treatment of mice with pancreatic cancer. NK4 was injected into mice between 3 and 28 days after the inoculation of human pancreatic cancer cells (SUIT-2). (B) Inhibition of primary tumor growth by NK4. Photographs show appearance of the primary pancreatic cancer. (C) Histological analysis of the effect of NK4-treatment on tumor angiogenesis (left) and apoptosis (right). NK4-treatment reduced the number of vessel numbers, while apoptotic death of cancers was enhanced by NK4. (D) Inhibitory effects of NK4 on peritoneal metastasis. Left: Typical macroscopic findings. Middle: Changes in the number of metstatic nodules. Right: Changes in the ascite volumes. (E) Prolonged survival of tumor-bearing mice treated with NK4.

long-term survival and a recurrence-free status. Targeting tumor angiogenesis and blockade of HGF-mediated invasion of cancer cells may prove to be potential therapy for patients with pancreatic cancer.

5.4 Therapy combining NK4 with other treatments

Anti-cancer chemotherapy is widely used for the suppression of malignant tumors with or without surgical treatment. Therapy regimens that combine anti-cancer chemo drugs and NK4 enhance their anti-tumor effect (Matsumoto *et al.*, 2011). Irradiation therapy often enhances cancer metastasis, especially in cases of pancreatic carcinoma, and this is associated with the irradiation-induced up-regulation of HGF in fibroblasts (Qian *et al.*, 2003; Ohuchida *et al.*, 2004). Thus, NK4 may overcome these irradiation-associated side effects.

Epidermal growth factor receptor (EGFR) kinase inhibitors, such as Gefitinib, are used to treat non-small cell lung cancers that have activating mutations in the EGFR gene, but most of these tumors become resistant to EGFR-kinase inhibitors due to enhancement of HGF-MET signals (Engelman *et al.*, 2007; Yano *et al.*, 2008; Okamoto *et al.*, 2010). Thus, NK4 treatment may reverse HGF-induced resistance to Gefitinib.

Endocrine Delivery System of NK4, an HGF-Antagonist and Anti-Angiogenic Regulator, for Inhibitions of Tumor
Growth, Invasion and Metastasis

129

Recently, it was demonstrated that NK4-mediated tumor regression depends on the infiltration of cytotoxic T lymphocytes (Kubota *et al.*, 2009b). Importantly, depletion of CD8+ cells markedly abrogated the anti-tumor activity of NK4 in a mouse model of colon cancer. NK4 enhances immune responses in dendritic cells *in vitro*. Thus, NK4 may also have utility for anti-tumor immunotherapy.

There is now ample evidence that NK4 is useful for the inhibition of growth, invasion and metastasis in various types of tumors, such as gastric carcinoma (Hirao *et al.*, 2002), pancreas cancer (Tomioka *et al.*, 2001), prostate cancer (Davies *et al.*, 2003), multiple myeloma (Du *et al.*, 2007) and melanoma (Kishi *et al.*, 2009) (**Table-1**). These results support our hypothesis that HGF is a key determinant of tumor malignancy (Matsumoto *et al.*, 1996b).

Tumor diseases (Cell lines and treatment)	NK4 therapy	Outcome	Literature
A. Digestive system: Gastric carcinoma (TMK1 cells, ip, Mouse)	Adeno-NK4, ip	Inhibitions of growth and metastasis, Anti-angiogenesis, Reduced ascites	Ueda K *et al.*, Eur J Cancer 40: 2135-2142 (2004)
Hepatic carcinoma (HUH7 cells, portal vein, Mouse)	Adeno-NK4, iv	Inhibitions of growth, Anti-angiogenesis, Prolonged survival	Son G *et al.*, J Hepatol 45: 688-695 (2006)
Gallbladder cancer (GB-d1 cells, sc, Mouse)	NK4, sc	Inhibitions of growth and invasion	Date K *et al.*, Oncogene 17: 3045-354 (1998)
Pancreatic carcinoma (SUIT-2 cells, intra-pancreas, Mouse)	r-NK4, ip	Inhibitions of growth, invasion and metastasis, Anti-angiogenesis, Reduced ascites, Prolonged survival	Tomioka D *et al.*, Cancer Res 61: 7518-7524 (2001)
Colon carcinoma (MC-38 cells, intra-spleen, Mouse)	NK4 cDNA, bolus iv (hydrodynamics)	Inhibitions of growth, invasion and metastasis, Anti-angiogenesis, Prolonged survival	Wen J *et al.*, Cancer Gen Ther 11: 419-430 (2004)
B. Respiratory system: Lung carcinoma (Lewis carcinoma, sc, Mouse)	r-NK4, sc	Inhibitions of growth and metastasis, Anti-angiogenesis, Enhanced apoptosis	Kuba K *et al.*, Cancer Res 60: 6737-6743 (2000)
Lung carcinoma (A549 cells, sc, Mouse)	Adeno-NK4, intra-tumor or ip	Inhibition of growth, Anti-angiogenesis	Maemondo M *et al.*, Mol Ther 5: 177-185 (2002)
Mesothelioma (EHMES-10 cells, sc, Mouse)	Adeno-NK4, intra-tumor	Inhibition of growth, Enhanced apoptosis, Anti-angiogenesis	Suzuki Y *et al.*, Int J Cancer 127: 1948-1957 (2010)

C. Reproductive system:			
Prostate carcinoma (PC-3 cells, sc, Mouse)	r-NK4, sc (osmotic pump)	Inhibition of growth, Anti-angiogenesis	Davies G et al., Int J Cancer 106: 348-354 (2003)
Ovarian carcinoma (HRA cells, ip, Mouse)	NK4 gene, Stable transfection	Inhibition of metastasis, Prolonged survival	Saga Y et al., Gene Ther 8: 1450-1455 (2001)
D. Hematopoietic system:			
Lymphoma (E.G7-OVA cells, sc, Mouse)	Adeno-NK4, intra-tumor (with DC)	Inhibition of growth, Anti-angiogenesis, Induction of CTL	Kikuchi T et al., Blood 100: 3950-3959 (2003)
Multiple myeloma (KMS11/34 cells, sc, Mouse)	Adeno-NK4, im	Inhibition of growth, Anti-angiogenesis, Enhanced apoptosis	Du W et al., Blood 109: 3042-3049 (2007)
E. Other organ or tissues:			
Melanoma (B16F10 cells, sc, Mouse)	Adeno-NK4, iv	Inhibitions of growth and metastasis, Anti-angiogenesis	Kishi Y et al., Cancer Sci 100: 1351-1358 (2009)
Glioblastoma (U-87 MG cells, Intra-brain, Mouse)	r-NK4, intra-tumor	Inhibition of growth, Anti-angiogenesis, Enhanced apoptosis	Brockmann MA et al., Clin Cancer Res 9: 4578-4585 (2003)
Breast carcinoma (MDAMB231 cells, sc, Mouse)	r-NK4, sc	Inhibition of growth, Anti-angiogenesis	Martin TA et al., Carcinogenesis 24: 1317-1323 (2003)

Adeno-NK4, adenoviral vector carrying NK4 cDNA; r-NK4, recombinant NK4 protein; sc, subcutaneous; iv, intravenous; ip, intraperitoneal; im, intramuscular; DC, dendritic cells; and CTL, cytotoxic T lymphocytes.

Table 1. Representative studies to show therapeutic effects of NK4 on distinct types of tumors in animal models

6. Hydodynamics-based NK4 gene therapy for colon cancer inhibition

Hydrodynamic delivery has emerged as the simplest and effective method for intracellular delivery of subjective genes in rodents; this process requires no special equipment. The system employs a physical force generated by the rapid injection of a large volume of solution into a blood vessel to enhance the permeability of endothelium and the plasma membrane of the parenchyma cells, such as hepatocytes, to facilitate a delivery of the substance into cells (Bonamassa et al., 2011). Using this technique in mice, we established an endocrine delivery system for NK4 that leads to an inhibition of the malignant behavior of cancers, as follows.

6.1 NK4 supplementation system via hydrodynamic gene delivery in mice

Numerous clinical studies have indicated the apparent increases in serum HGF levels in patients during the progression of cancers (Wu *et al.*, 1998; Osada *et al.*, 2008). It is likely that HGF in blood protects cancer cell suspension from anoikis-like cell death (Zeng *et al.*, 2002). Thus, we predict that over-production of NK4 in blood would overcome the HGF-mediated metastatic events seen in blood flow (and possibly in local sites). Hydrodynamic-based gene delivery is known to achieve an efficient expression of exogenous genes predominantly in the liver but much lesser in the kidney and spleen (Suda *et al.*, 2007). Based on this background, we established a method for the induction and maintenance of higher levels of NK4 in blood through repeated injections of NK4 cDNA-containing plasmid.

For hydrodynamic-based gene delivery, 5 microgram of plasmid DNA (pCAGGS-NK4), or pCAGGS-empty (as a control), in saline was injected within 5 seconds into tail veins of mice at 2.4 ml per 30g body weight (Wen *et al.*, 2004; 2007). As a result, exogenous NK4 was detected, and plasma NK4 reached a mean value of 49.5 ng/ml 24 hours post-bolus injection and decreased to 15.4 ng/ml on day 3. Following the second and third injections, the plasma NK4 level again reached approximately 70 and 130 ng/ml on days 8 and 15, respectively. Thus, plasma NK4 levels increased following additional administration of the expression plasmid, and were maintained at levels of > 8 ng/ml during 3 weeks post-treatment (**Fig. 5**).

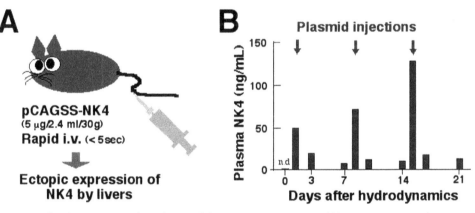

Fig. 5. Hydrodynamics-mediated NK4 delivery system in mice. (A) An experimental protocol of NK4 gene administration. Five microgram of pCAGGS-NK4 was administered intravenously into mice on day 0, 7 and 14. (B) Changes in plasma NK4 levels following repetitive administration of expression plasmid for NK4. Arrows mean the time of plasmid administration. See reference (Wen *et al.*, 2007) for further information.

6.2 Inhibition of colon cancer metastasis by NK4 gene delivery

Colon cancer is one of the most common cancers in the world, with a high propensity to metastasize: 30-40% of patients have metastatic disease at the initial diagnosis. The liver is the most frequent site of metastasis, and hepatic failure is a lethal event during colon cancer. Thus, direct inhibition of the dissociation, spreading and invasion of cancer cells is expected to become efficient treatment. With regard to this, HGF stimulates the invasion of MC-38 mouse colon cancer cells across MatriGel (Parr *et al.*, 2000), which is composed of laminin and other matrices and mimics the basement membrane *in vivo*. In this model, NK4 has inhibited the HGF-mediated migration of MC-38 cells in a culture model of colon cancer invasion. This anti-invasive effect of NK4, obtained by *in vitro* studies, is demonstrated *in vivo* in the following two studies.

An hepatic metastatic model was prepared by the injection of mouse MC-38 cells into the spleen. During the progression of colon cancer in hepatic tissues, HGF was over-produced by hepatic sinusoidal cells, while MET tyrosine phosphorylation became evident, particularly around the front lines of invasive zones. Supplementation of NK4 in blood and livers via a single injection of NK4-cDNA containing plasmid (pCAGGS-NK4) resulted in the loss of MET tyrosine phosphorylation (**Fig. 6**). Under such a MET-inactivated condition by NK4 treatment, hepatic invasion by colon carcinoma was strongly inhibited (Wen *et al.*, 2004).

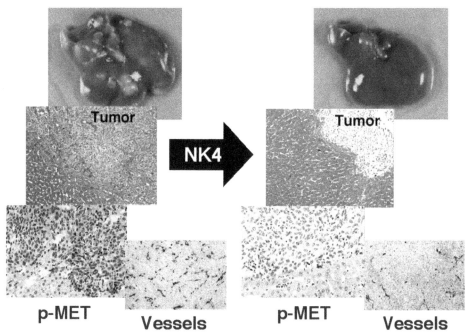

Fig. 6. Successful outcome of hydrodynamics-based NK4 gene therapy in a mouse model of colon cancer. The hepatic invasion model is prepared by intra-splenic inoculation of MC-38 colon carcinoma in mice. In the control group, invasion of carcinoma cells into neighboring hepatic areas becomes evident, along with an induction of MET tyrosine phosphorylation (p-MET) and an increase in vessel numbers. In contrast, NK4 suppresses tumor invasion by inhibiting MET tyrosine phosphorylation and reducing angiogenesis. As a result, NK4 gene therapy prolongs the survival of these mice (Wen *et al.*, 2004).

Repeated administrations of NK4-containing plasmid DNA also inhibited the malignant behaviors of colon carcinoma (Wen *et al.*, 2007). Actually, NK4 repetitive gene therapy potently inhibited the muscular invasion of MC-38 carcinoma cells. Furthermore, angiogenesis in the colon cancer was markedly suppressed by NK4 repetitive therapy, along with an increase in tumor apoptosis. Overall, the number of hepatic metastatic nodules was dramatically decreased by the repeated injections of NK4-cDNA containing plasmid. This study provides an anti-tumor model where NK4 is supplemented via a hydrpdynamics-based gene therapy (Wen *et al.*, 2007).

Recently, hydrodynamic gene delivery using a rapid injection of a relatively large volume of DNA solution has facilitated experimental gene therapy studies, particularly in rodents (Suda *et al.*, 2007). This method is superior to the existing delivery systems because of its simplicity, efficiency, and versatility. Hydrodynamic gene delivery is also useful for supplementation of HGF, an intrinsic repair factor, for the inhibition of, or recovery from, intractable organ diseases, such as acute renal failure (Dai *et al.*, 2002) or pulmonary airway hyper-responsiveness during asthma (Okunishi *et al.*, 2005). In these experiments, plasma HGF levels were sustained within a pharmacological range (3-30 ng/ml). Wide success in applying hydrodynamic principles to delivery of NK4- or HGF-related DNA, RNA, proteins, and synthetic compounds, into the cells in various tissues of small animals, has inspired the recent attempts at establishing a hydrodynamic procedure for clinical use.

7. Summary and perspective

NK4-related studies provided a proof-of-concept that MET signaling from stroma-derived HGF plays a pivotal role in eliciting tumor invasion and metastasis (Matsumoto & Nakamura., 2005; Nakamura *et al.*, 2010). Human genetic studies also strengthened the important role of MET activation for tumor malignancy. There is now ample evidence to demonstrate the role of MET mutations in tumor malignancy (Lengyel *et al.*, 2007; Matsumoto *et al.*, 2008a,b; Pao *et al.*, 2011). Of interest, mutation of the von-Hippel-Lindau (VHL) gene leads to renal clear cell carcinoma through constitutive MET tyrosine phosphorylation (Nakaigawa *et al.*, 2006), hence suggesting a critical role of wild-type VHL in inhibiting MET over-activation as a negative regulator.

During the progression of malignant tumors, soluble MET is producible by carcinoma cells through an ectodomain shedding cascade (Wader *et al.*, 2011). Soluble MET inhibits the HGF-MET complex and signaling transduction. Thus, MET shedding system is considered as a self-defense response that minimizes tumor metastasis. Likewise, an NK4-like fragment of the HGF α-chain can be secreted from human breast carcinoma, which inhibits MET tyrosine phosphorylation (Wright *et al.*, 2009). Thus, "endogenous" soluble MET and NK4-like variant appear to reduce HGF-MET signaling and delay tumor progression, but this response is insufficient, allowing for tumor metastasis. Thus, supplemental therapy with NK4 is a reasonable strategy to completely block tumor metastasis.

The hope is that angiogenesis inhibition might control tumor metastasis (Yancopoulos *et al.*, 1998). However, long-term use of angiogenesis inhibitors, such as VEGF inhibitor, results in hypoxia-resistance (Fischer *et al.*, 2007), possibly due to hypoxia-induced MET up-regulation by cancer (Bottaro & Liotta, 2003). NK4 is an angiogenesis inhibitor with the ability to inhibit MET activation, and discovery of this fragment opened up a new avenue for the development of freeze-and-dormancy therapy (**Fig. 7**). Thus, NK4 is now defined as "Malignostatin". In addition to NK4, several anti-metastatic drugs have been proposed, with a major focus on small molecules that inhibit the tyrosine kinase activity of MET; ribozyme; small-interfering RNA; anti-HGF antibodies; soluble MET; and HGF-variant decoys (Jiang *et al.*, 2005; Benvenuti & Comoglio, 2007; Eder *et al.*, 2009; Underiner *et al.*, 2010; Cecchi *et al.*, 2010). HGF-MET targeting research will shed more light on cancer biology, pathology and new technologies to overcome host death due to cancer metastasis.

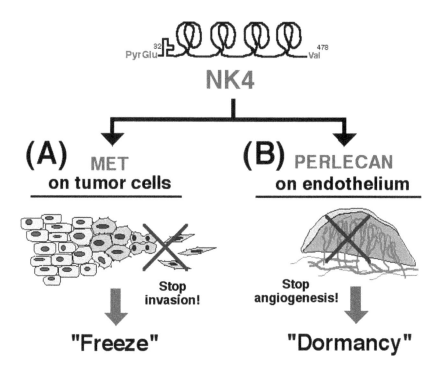

Fig. 7. Freeze-and-dormancy therapy of malignant tumors by NK4/malignostatin. NK4 blocks tumor invasion and metastasis through an inhibition of HGF-MET signals as an HGF-antagonist. Furthermore, NK4 inhibits tumor angiogenesis via a perlecan-dependent mechanism. Such a dual function of NK4 contributes to "freeze" and "dormancy" anti-cancer therapy.

8. Acknowledgement

The publication of this manuscript was supported by grants from the Ministry of Education, Science, Technology, Sports and Culture of Japan (the 21st Century global COE program to TN). We would like to thank all co-workers for the continuous studies on the roles of HGF and NK4 during cancer progression. We are also grateful to James L. McDonald (Scientific Editorial Services; Harrison, AR) for language assistance.

9. References

Benvenuti, S. & Comoglio, P. M. (2007). The MET receptor tyrosine kinase in invasion and metastasis. *Journal of Cellular Physiology*, Vol.213, No.2, (November 2009), pp. 316-325, ISSN 0021-9541

Beppu, K.; Uchiyama, A.; Morisaki, T.; Matsumoto, K.; Nakamura, T.; Tanaka, M. & Katano, M. (2001). Hepatocyte growth factor production by peripheral blood mononuclear cells of recurrent cancer patients. *Anticancer Research*, Vol.21, No.3C, (May-June 2001), pp. 2195-2200, ISSN 0250-7005.

Bhowmick, N. A.; Chytil, A.; Plieth, D.; Gorska, A. E.; Dumont, N.; Shappell, S.; Washington, M. K.; Neilson, E. G. & Moses, H. L. (2004). TGF-beta signaling in fibroblasts modulates the oncogenic potential of adjacent epithelia. *Science*, Vol.303, No.5659, (February 2004), pp. 848-851, ISSN 0036-8075

Birchmeier, C. & Gherardi, E. (1998). Developmental roles of HGF/SF and its receptor, the c-Met tyrosine kinase. *Trends in Cell Biology*, Vol.8, No.10, (October 1998), pp. 404-410, ISSN: 0962-8924.

Bonamassa, B.; Hai, L. & Liu, D. (2011) Hydrodynamic gene delivery and its applications in pharmacological research. *Pharmaceutical Research*, Vol.28, No.4, (April 2011), pp. 694-701, ISSN 0724-8741

Bottaro, D. P.; Rubin, J. S.; Faletto, D. L.; Chan, A. M.; Kmiecik, T. E.; Vande Woude, G. F. & Aaronson, S. A. (1991). Identification of the hepatocyte growth factor receptor as the c-met proto-oncogene product. *Science*, Vol.251, No.4995, (February 1991) pp. 802-804, ISSN 0036-8075

Bottaro, D. P. & Liotta, L. (2003). Cancer: Out of air is not out of action. *Nature*, Vol.423, No.6940, (June 2003), 593-595, ISSN 0028-0836

Bussolino, F.; Di Renzo, M. F.; Ziche, M.; Bocchietto, E.; Olivero, M.; Naldini, L.; Gaudino, G.; Tamagnone, L.; Coffer, A. & Comoglio, P. M. (1992). Hepatocyte growth factor is a potent angiogenic factor which stimulates endothelial cell motility and growth. *The Journal of Cell Biology*, Vol.119, No.3, (November 1992), pp. 629-641, ISSN 0021-9525

Cecchi, F.; Rabe, D. C. & Bottaro, D. P. (2010). Targeting the HGF/Met signalling pathway in cancer. *European Journal of Cancer*, Vol.46, No.7, (May 2010), pp. 1260-1270, ISSN 959-8049

Cooper, C. S.; Park, M.; Blair, D. G.; Tainsky, M. A.; Huebner, K.; Croce, C. M. & Vande Woude, G. F. (1984). Molecular cloning of a new transforming gene from a chemically transformed human cell line. *Nature*, Vol.311, No.5981, (September 1984), pp. 29-33, ISSN 0028-0836

Dai, C.; Yang, J. & Liu, Y. (2002). Single injection of naked plasmid encoding hepatocyte growth factor prevents cell death and ameliorates acute renal failure in mice. *Journal of the American Society of Nephrology*, Vol.13, No.2, (February 2002) pp. 411-422, ISSN 1046-6673

Date, K.; Matsumoto, K.; Shimura, H.; Tanaka, M. & Nakamura, T. (1997). HGF/NK4 is a specific antagonist for pleiotrophic actions of hepatocyte growth factor. *FEBS Letter*, Vol.420, No.1, (December 1997), pp. 1-6, ISSN 0014-5793

Date, K.; Matsumoto, K.; Kuba, K.; Shimura, H.; Tanaka, M. & Nakamura, T. (1998). Inhibition of tumor growth and invasion by a four-kringle antagonist (HGF/NK4) for hepatocyte growth factor. *Oncogene*, Vol.17, No.23, (December 1998), pp. 3045-3054, ISSN 0950-9232

Davies, G.; Mason, M. D.; Martin, T. A.; Parr, C.; Watkins, G.; Lane, J.; Matsumoto, K.; Nakamura, T. & Jiang, W. G. (2003). The HGF/SF antagonist NK4 reverses fibroblast- and HGF-induced prostate tumor growth and angiogenesis in vivo. *International Journal of Cancer*, Vol.106, No.3, (September 2003), pp. 348-354, ISSN 0020-7136

Dean, M.; Park, M.; Le Beau, M. M.; Robins, T. S.; Diaz, M. O.; Rowley, J. D.; Blair, D. G. & Vande Woude, G. F. (1985). The human met oncogene is related to the tyrosine kinase oncogenes. *Nature.* Vol.318, No.6044, (November-December 1985), pp. 385-388, ISSN 0028-0836

Du, W.; Hattori, Y.; Yamada, T.; Matsumoto, K.; Nakamura, T.; Sagawa, M.; Otsuki, T.; Niikura, T.; Nukiwa, T. & Ikeda, Y. (2007). NK4, an antagonist of hepatocyte growth factor (HGF), inhibits growth of multiple myeloma cells: molecular targeting of angiogenic growth factor. *Blood*, Vol.109, No.7, (April 2007), pp. 3042-3049, ISSN 0006-4971

Eder, J. P.; Vande Woude, G. F.; Boerner, S. A. & LoRusso, P. M. (2009). Novel therapeutic inhibitors of the c-Met signaling pathway in cancer. *Clinical Cancer Research*, Vol.15, No.7, (April 2009), pp. 2207-2214, ISSN 1078-0432

Engelman, J. A.; Zejnullahu, K.; Mitsudomi, T.; Song, Y.; Hyland, C.; Park, J. O.; Lindeman, N.; Gale, C. M.; Zhao, X.; Christensen, J.; Kosaka, T.; Holmes, A. J.; Rogers, A. M.; Cappuzzo, F.; Mok, T.; Lee, C.; Johnson, B. E.; Cantley, L. C. & Jänne, P. A. (2007). MET amplification leads to gefitinib resistance in lung cancer by activating ERBB3 signaling. *Science*, Vol.316, No.5827, (May 2007), pp. 1039-1043, ISSN 0036-8075

Ferlay, J.; Shin, H. R.; Bray, F.; Forman, D.; Mathers, C. & Parkin, D. M. (2010). Estimates of worldwide burden of cancer in 2008: GLOBOCAN 2008. *International Journal of Cancer*, Vol.127, No.12, (December 2010), pp. 2893-2917, ISSN 0020-7136

Fischer, C.; Jonckx, B.; Mazzone, M.; Zacchigna, S.; Loges, S.; Pattarini, L.; Chorianopoulos, E.; Liesenborghs, L.; Koch, M.; De Mol, M.; Autiero, M.; Wyns, S.; Plaisance, S.; Moons, L.; van Rooijen, N.; Giacca, M.; Stassen, J. M.; Dewerchin, M.; Collen, D. & Carmeliet, P. (2007). Anti-PlGF inhibits growth of VEGF(R)-inhibitor-resistant tumors without affecting healthy vessels. *Cell*, Vol.131, No.3, (November 2007), pp. 463-475, ISSN 0092-8674

Franco, M.; Muratori, C.; Corso, S.; Tenaglia, E.; Bertotti, A.; Capparuccia, L.; Trusolino, L.; Comoglio, P. M. & Tamagnone, L. (2010). The tetraspanin CD151 is required for Met-dependent signaling and tumor cell growth. *Journal of Biological Chemistry*, Vol.285, No.50, (December 2010), pp. 38756-38764, ISSN 0021-9258

Geiger, T. R. & Peeper, D. S. (2009). Metastasis mechanisms. *Biochimica et Biophysica Acta*, Vol.1796, No.2, (August 2009), pp. 293-308, ISSN 0006-3002

Grugan, K. D.; Miller, C. G.; Yao, Y.; Michaylira, C. Z.; Ohashi, S.; Klein-Szanto, A. J.; Diehl, J. A.; Herlyn, M.; Han, M.; Nakagawa, H. & Rustgi, A. K. (2010). Fibroblast-secreted hepatocyte growth factor plays a functional role in esophageal squamous cell carcinoma invasion. *Proceeding of National Academy of Sciences*, Vol.107, No.24, (June 2010), pp. 11026-11031, ISSN 0027-8424

Hasina, R.; Matsumoto, K.; Matsumoto-Taniura, N.; Kato, I.; Sakuda, M. & Nakamura, T. (1999). Autocrine and paracrine motility factors and their involvement in invasiveness in a human oral carcinoma cell line. *British Journal of Cancer*, Vol.80, No.11, (August 1999), pp. 1708-1717, ISSN 0007-0920

Higuchi, O.; Mizuno, K.; Vande Woude, G. F. & Nakamura, T. (1992). Expression of c-met proto-oncogene in COS cells induces the signal transducing high-affinity receptor for hepatocyte growth factor. *FEBS Letter*, Vol.301, No.3, (April 1999), pp. 282-286, ISSN 0014-5793

Hiscox, S. & Jiang, W. G. (1999). Hepatocyte growth factor/scatter factor disrupts epithelial tumour cell-cell adhesion: involvement of beta-catenin. *Anticancer Research*, Vol.19, No.1A, (January February, 1999), pp. 509-517, ISSN 0250-7005.

Hiscox, S.; Parr, C.; Nakamura, T.; Matsumoto, K.; Mansel, R. E. & Jiang, W. G. (2000). Inhibition of HGF/SF-induced breast cancer cell motility and invasion by the

HGF/SF variant, NK4. *Breast Cancer Research and Treatment*, Vol.59, No.3, (February 2000), pp. 245-254, ISSN 0167-6806

Hirao, S.; Yamada, Y.; Koyama, F.; Fujimoto, H.; Takahama, Y.; Ueno, M.; Kamada, K.; Mizuno, T.; Maemondo, M.; Nukiwa, T.; Matsumoto, K.; Nakamura, T. & Nakajima, Y. (2002). Tumor suppression effect using NK4, a molecule acting as an antagonist of HGF, on human gastric carcinomas. *Cancer Gene Therapy*, Vol.9, No.8, (August 2002), pp. 700-707, ISSN 0929-1903

Jiang, W. G.; Martin, T. A.; Matsumoto, K.; Nakamura, T. & Mansel, R. E. (1999a). Hepatocyte growth factor/scatter factor decreases the expression of occludin and transendothelial resistance (TER) and increases paracellular permeability in human vascular endothelial cells. *Journal of Cellular Physiology*, Vol.181, No.2, (November 1999), pp. 319-329, ISSN 0021-9541

Jiang, W. G.; Hiscox, S. E.; Parr, C.; Martin, T. A.; Matsumoto, K.; Nakamura, T. & Mansel, R. E. (1999b) Antagonistic effect of NK4, a novel hepatocyte growth factor variant, on in vitro angiogenesis of human vascular endothelial cells. *Clinical Cancer Research*, Vol.5, No.11, (November 1999), pp. 3695-3703, ISSN 1078-0432

Jiang, Y.; Xu, W.; Lu, J.; He, F. & Yang, X. (2001). Invasiveness of hepatocellular carcinoma cell lines: contribution of hepatocyte growth factor, c-met, and transcription factor Ets-1. *Biochemical and Biophysical Research Communications*, Vol.286, No.5, (September, 2001), pp. 1123-1130, ISSN 0006-291X

Jiang, W. G.; Martin, T. A.; Parr, C.; Davies, G.; Matsumoto, K. & Nakamura, T. (2005). Hepatocyte growth factor, its receptor, and their potential value in cancer therapies. *Critical Reviews in Oncology/Hematology*, Vol.53, No.1, (January 2005), pp. 35-69, ISSN 1040-8428

Kajiya, K.; Hirakawa, S.; Ma, B.; Drinnenberg, I. & Detmar, M. (2005). Hepatocyte growth factor promotes lymphatic vessel formation and function. *EMBO Journal*, Vol.24, No.16, (August 2005), pp. 2885-2895, ISSN 0261-4189

Kanayama, S.; Yamada, Y.; Kawaguchi, R.; Tsuji, Y.; Haruta, S. & Kobayashi, H. (2008). Hepatocyte growth factor induces anoikis resistance by up-regulation of cyclooxygenase-2 expression in uterine endometrial cancer cells. *Oncology Reports*, Vol.19, No.1, (January 2008), pp. 117-122, ISSN 1021-335X

Kimura, T.; Sakisaka, T.; Baba, T.; Yamada, T. & Takai Y. (2006). Involvement of the Ras-Ras-activated Rab5 guanine nucleotide exchange factor RIN2-Rab5 pathway in the hepatocyte growth factor-induced endocytosis of E-cadherin. *Journal of Biological Chemistry*, Vol.281, No.15, (April 2006), pp. 10598-10609, ISSN 0021-9258

Kishi, Y.; Kuba, K.; Nakamura, T.; Wen, J.; Suzuki, Y.; Mizuno, S.; Nukiwa, T.; Matsumoto, K. & Nakamura, T. (2009). Systemic NK4 gene therapy inhibits tumor growth and metastasis of melanoma and lung carcinoma in syngeneic mouse tumor models. *Cancer Science*, Vol.100, No.7, (July 2009) pp. 1351-1358, ISSN 1347-9032

Kuba, K.; Matsumoto, K.; Date, K.; Shimura, H.; Tanaka, M. & Nakamura, T. (2000) HGF/NK4, a four-kringle antagonist of hepatocyte growth factor, is an angiogenesis inhibitor that suppresses tumor growth and metastasis in mice. *Cancer Research*, Vol.60, No.23, (December 2000), pp. 6737-6743, ISSN 0008-5472

Kubota, T.; Taiyoh, H.; Matsumura, A.; Murayama, Y.; Ichikawa, D.; Okamoto, K.; Fujiwara, H.; Ikoma, H.; Nakanishi, M.; Kikuchi, S.; Sakakura, C.; Ochiai, T.; Kokuba, Y.; Taniguchi, H.; Sonoyama, T.; Matsumoto, K.; Nakamura, T. & Otsuji. E. (2009a). NK4, an HGF antagonist, prevents hematogenous pulmonary metastasis by inhibiting adhesion of CT26 cells to endothelial cells. *Clinical and Experimental Metastasis*, Vol.26, No.5, (May 2009), pp. 447-456, ISSN 0262-0898

Kubota, T.; Taiyoh, H.; Matsumura, A.; Murayama, Y.; Ichikawa, D.; Okamoto, K.; Fujiwara, H.; Ikoma, H.; Nakanishi, M.; Kikuchi, S.; Ochiai, T.; Sakakura, C.; Kokuba, Y.; Sonoyama, T.; Suzuki, Y.; Matsumoto, K.; Nakamura, T. & Otsuji, E. (2009b). Gene transfer of NK4, an angiogenesis inhibitor, induces CT26 tumor regression via tumor-specific T lymphocyte activation. *International Journal of Cancer*, Vol.125, No.12, (December 2009), pp. 2879-2886, ISSN 0020-7136

Konishi, T.; Takehara, T.; Tsuji, T.; Ohsato, K.; Matsumoto, K. & Nakamura, T. (1991). Scatter factor from human embryonic lung fibroblasts is probably identical to hepatocyte growth factor. *Biochemical and Biophysical Research Communications*, Vol.180, No.2, (October, 1991), pp. 765-773, ISSN 0006-291X

Laterra, J.; Nam, M.; Rosen, E.; Rao, J. S.; Lamszus, K.; Goldberg, I. D. & Johnston, P. (1997). Scatter factor/hepatocyte growth factor gene transfer enhances glioma growth and angiogenesis in vivo. *Laboratory Investigation*, Vol.76, No.4, (April 1997), pp. 565-577, ISSN 0023-6837

Lengyel, E.; Sawada, K. & Salgia, R. (2007). Tyrosine kinase mutations in human cancer. *Current Molecular Medicine*, Vol.7, No.1, (February 2007), pp. 77-84, ISSN 1566-5240

Li, H.; Shimura, H.; Aoki, Y.; Date, K.; Matsumoto, K.; Nakamura, T. & Tanaka, M. (1998). Hepatocyte growth factor stimulates the invasion of gallbladder carcinoma cell lines in vitro. *Clinical and Experimental Metastasis*, Vol.16, No.1, (January 1998), pp. 74-82, ISSN 0262-0898

Maehara, N.; Matsumoto, K.; Kuba, K.; Mizumoto, K.; Tanaka, M. & Nakamura, T. (2001). NK4, a four-kringle antagonist of HGF, inhibits spreading and invasion of human pancreatic cancer cells. *British Journal of Cancer*, Vol.84, No.6, (March 2001), pp. 864-873, ISSN 0007-0920

Matsumoto, K.; Tajima, H.; Okazaki, H. & Nakamura, T. (1992). Negative regulation of hepatocyte growth factor gene expression in human lung fibroblasts and leukemic cells by transforming growth factor-beta1 and glucocorticoids. *Journal of Biological Chemistry*, Vol.267, No.35, (December 1992), pp. 24917-24920, ISSN 0021-9258

Matsumoto, K.; Matsumoto, K.; Nakamura, T. & Kramer. R. H. (1994). Hepatocyte growth factor/scatter factor induces tyrosine phosphorylation of focal adhesion kinase (p125FAK) and promotes migration and invasion by oral squamous cell carcinoma cells. *Journal of Biological Chemistry*, Vol.269, No.50, (December 1994), pp. 31807-31813, ISSN 0021-9258

Matsumoto, K.; Date, K.; Shimura, H. & Nakamura, T. (1996a). Acquisition of invasive phenotype in gallbladder cancer cells via mutual interaction of stromal fibroblasts and cancer cells as mediated by hepatocyte growth factor. *Japanese Journal of Cancer Research*, Vol.87, No.7, (July 1996), pp. 702-710, ISSN 0910-5050

Matsumoto, K.; Date, K.; Ohmichi, H. & Nakamura, T. (1996b). Hepatocyte growth factor in lung morphogenesis and tumor invasion: role as a mediator in epithelium-mesenchyme and tumor-stroma interactions. *Cancer Chemotherapy and Pharmacology*, Vol.38 (Supplement), pp. S42-S47, ISSN 0344-5704

Matsumoto, K.; Kataoka, H.; Date, K. & Nakamura, T. (1998). Cooperative interaction between alpha- and beta-chains of hepatocyte growth factor on c-Met receptor confers ligand-induced receptor tyrosine phosphorylation and multiple biological responses. *Journal of Biological Chemistry*, Vol.273, No.36, (September 1998), pp. 22913-22920, ISSN 0021-9258

Matsumoto-Taniura, N.; Matsumoto, K. & Nakamura, T. (1999). Prostaglandin production in mouse mammary tumour cells confers invasive growth potential by inducing hepatocyte growth factor in stromal fibroblasts. *British Journal of Cancer*, Vol.81, No.2, (September 1999), pp. 194-202, ISSN 0007-0920

Matsumoto, K. & Nakamura, T. (2003). NK4 (HGF-antagonist/angiogenesis inhibitor) in cancer biology and therapeutics. *Cancer Science*, Vol.94, No.4, (April 2003), pp. 321-327, ISSN 1347-9032

Matsumoto, K. & Nakamura, T. (2005). Mechanisms and significance of bifunctional NK4 in cancer treatment. *Biochemical and Biophysical Research Communications*, Vol.333, No.2, (July 2005), pp. 316-327, ISSN 0006-291X

Matsumoto, K. & Nakamura, T. (2006). Hepatocyte growth factor and the Met system as a mediator of tumor-stromal interactions. *International Journal of Cancer*, Vol.119, No.3, (August, 2006), pp. 477-483, ISSN 0020-7136

Matsumoto, K. & Nakamura, T. (2008a). NK4 gene therapy targeting HGF-Met and angiogenesis. *Frontiers in Bioscience*, Vol.13, No.1, (January 2008), pp. 1943-1951, ISSN 1093-9946

Matsumoto, K.; Nakamura, T.; Sakai, K. & Nakamura, T. (2008b). Hepatocyte growth factor and Met in tumor biology and therapeutic approach with NK4. *Proteomics*, Vol.8, No.16, (August 2008), pp. 3360-3370, ISSN 1615-9861

Matsumoto, G.; Omi, Y.; Lee, U.; Kubota, E. & Tabata, Y. (2011). NK4 gene therapy combined with cisplatin inhibits tumour growth and metastasis of squamous cell carcinoma. *Anticancer Research*, Vol.31, No.1, (January 2011), pp. 105-111, ISSN 0250-7005

Miura, H.; Nishimura, K.; Tsujimura, A.; Matsumiya, K.; Matsumoto, K.; Nakamura, T. & Okuyama, A. (2001). Effects of hepatocyte growth factor on E-cadherin-mediated cell-cell adhesion in DU145 prostate cancer cells. *Urology*, Vol.58, No.6, (December 2001), pp. 1064-1069, ISSN 0090-4295

Mueller, M. M. & Fusenig, N. E. (2004). Friends or foes - bipolar effects of the tumour stroma in cancer. *Nature Reviews Cancer*, Vol.4, No.11, (November 2004), pp. 839-849, ISSN 1474-175X

Nakabayashi, M.; Morishita, R.; Nakagami, H.; Kuba, K.; Matsumoto, K.; Nakamura, T.; Tano, Y. & Kaneda, Y. (2003). HGF/NK4 inhibited VEGF-induced angiogenesis in in vitro cultured endothelial cells and in vivo rabbit model. *Diabetologia*, Vol.46, No.1, (January 2003), pp. 115-123, ISSN 0012-186X

Nagakawa, O.; Murakami, K.; Yamaura, T.; Fujiuchi, Y.; Murata, J.; Fuse, H. & Saiki, I. (2000). Expression of membrane-type 1 matrix metalloproteinase (MT1-MMP) on prostate cancer cell lines. *Cancer Letter*, Vol.155, No.2, (July 2000), pp. 173-179, ISSN 0304-3835

Nakaigawa, N.; Yao, M.; Baba, M.; Kato, S.; Kishida, T.; Hattori, K.; Nagashima. Y. & Kubota, Y. (2006). Inactivation of von Hippel-Lindau gene induces constitutive phosphorylation of MET protein in clear cell renal carcinoma. *Cancer Research*, Vol.66, No.7, (April 2006), pp. 3699-3705, ISSN 0008-5472

Nakamura, T.; Nawa, K. & Ichihara, A. (1984). Partial purification and characterization of hepatocyte growth factor from serum of hepatectomized rats. *Biochemical and Biophysical Research Communications*, Vol.122, No.3, (August 1984), pp. 1450-1459, ISSN 0006-291X

Nakamura, T.; Nishizawa, T.; Hagiya, M.; Seki, T.; Shimonishi, M.; Sugimura, A.; Tashiro, K. & Shimizu, S. (1989). Molecular cloning and expression of human hepatocyte growth factor. *Nature*, Vol.342, No.6248, (November 1989), pp. 440-443, ISSN 0028-0836

Nakamura, T. (1991). Structure and function of hepatocyte growth factor. *Progress in Growth Factor Research*, Vol.3, No.1. pp. 67-85, ISSN: 0955-2235

Nakamura, T.; Matsumoto, K.; Kiritoshi, A.; Tano, Y. & Nakamura, T. (1997) Induction of hepatocyte growth factor in fibroblasts by tumor-derived factors affects invasive growth of tumor cells: in vitro analysis of tumor-stromal interactions. *Cancer Research*, Vol.57, No.15, (August 1997), pp. 3305-3313, ISSN 0008-5472

Nakamura, T. & Mizuno, S. (2010). The discovery of hepatocyte growth factor (HGF) and its significance for cell biology, life sciences and clinical medicine. *Proceedings of the Japan Academy Series B*, Vol.86, No.6, (June 2010), pp. 588-610 (2010), ISSN 0386-2208

Nakamura, T.; Sakai, K.; Nakamura, T. & Matsumoto, K. (2010). Anti-cancer approach with NK4: Bivalent action and mechanisms. *Anti-Cancer Agents in Medicinal Chemistry*, Vol.10, No.1, (January 2010), pp. 36-46, ISSN 1871-5206

Nakamura, Y.; Morishita, R.; Higaki, J.; Kida, I.; Aoki, M.; Moriguchi, A.; Yamada, K.; Hayashi, S.; Yo, Y.; Nakano, H.; Matsumoto, K.; Nakamura, T. & Ogihara, T. (1996). Hepatocyte growth factor is a novel member of the endothelium-specific growth factors: additive stimulatory effect of hepatocyte growth factor with basic fibroblast growth factor but not with vascular endothelial growth factor. *Journal of Hypertension*, Vol.14, No.9, (September 1996), pp. 1067-1072, ISSN 0263-6352

Ohuchida, K.; Mizumoto, K.; Murakami, M.; Qian, L. W.; Sato, N.; Nagai, E.; Matsumoto, K.; Nakamura, T. & Tanaka, M. (2004) Radiation to stromal fibroblasts increases invasiveness of pancreatic cancer cells through tumor-stromal interactions. *Cancer Research*, Vol.64, No.9, (May 2004), pp. 3215-3222, ISSN 0008-5472

Okamoto, W.; Okamoto, I.; Tanaka, K.; Hatashita, E.; Yamada, Y.; Kuwata, K.; Yamaguchi, H.; Arao, T.; Nishio, K.; Fukuoka, M.; Jänne, P. A. & Nakagawa, K. (2010). TAK-701, a humanized monoclonal antibody to hepatocyte growth factor, reverses gefitinib resistance induced by tumor-derived HGF in non-small cell lung cancer with an EGFR mutation. *Molecular Cancer Therapeutics*, 2010 Oct; Vol.9, No.10, (Octorber 2010), pp. 2785-92, ISSN 1535-7163

Okunishi, K.; Dohi, M.; Nakagome, K.; Tanaka, R.; Mizuno, S.; Matsumoto, K.; Miyazaki, J.; Nakamura, T. & Yamamoto, K. (2005) A novel role of hepatocyte growth factor as an immune regulator through suppressing dendritic cell function. *The Journal of Immunology*, Vol.175, No.7, (October 2005), pp. 4745-4753, ISSN 0022-1767

Olumi, A. F.; Grossfeld, G. D.; Hayward, S. W.; Carroll, P. R.; Tlsty, T.D. & Cunha, G. R. (1999). Carcinoma associated fibroblasts direct tumor progression of initiated human prostatic epithelium. *Cancer Research*, Vol.59, No.19, (October 1999), pp. 5002-5011, ISSN 0008-5472

Osada, S.; Kanematsu, M.; Imai, H. & Goshima, S. (2008). Clinical significance of serum HGF and c-Met expression in tumor tissue for evaluation of properties and treatment of hepatocellular carcinoma. *Hepatogastroenterology*, Vol.55, No.82-83, (March-April 2008), pp. 544-549, ISSN 0172-6390

Pai, R.; Nakamura, T.; Moon, W. S. & Tarnawski, A. S. (2003). Prostaglandins promote colon cancer cell invasion; signaling by cross-talk between two distinct growth factor receptors. *FASEB Journal*, Vol.17, No.12, (September 2003), pp. 1640-1647, ISSN 0892-6638

Parr, C.; Hiscox, S.; Nakamura, T.; Matsumoto, K. & Jiang, W. G. (2000). NK4, a new
HGF/SF variant, is an antagonist to the influence of HGF/SF on the motility and
invasion of colon cancer cells. *International Journal of Cancer*, Vol.85, No.4, (February
2000), pp. 563-570, ISSN 0020-7136

Parr, C.; Davies, G.; Nakamura, T.; Matsumoto, K.; Mason, M. D. & Jiang, W. G. (2001). The
HGF/SF-induced phosphorylation of paxillin, matrix adhesion, and invasion of
prostate cancer cells were suppressed by NK4, an HGF/SF variant. *Biochemical and
Biophysical Research Communications*, Vol.285, No.5, (August 2010), pp. 1330-1337,
ISSN 0006-291X

Pao, W. & Girard, N. (2011). New driver mutations in non-small-cell lung cancer. *The Lancet
Oncology*, Vol.12, No.2, (February 2011), pp. 175-180, ISSN 1470-2045

Qian, L. W.; Mizumoto, K.; Inadome, N.; Nagai, E.; Sato, N.; Matsumoto, K.; Nakamura, T. &
Tanaka, M. (2003). Radiation stimulates HGF receptor/c-Met expression that leads to
amplifying cellular response to HGF stimulation via upregulated receptor tyrosine
phosphorylation and MAP kinase activity in pancreatic cancer cells. *International
Journal of Cancer*, Vol.104, No.5, (May 2003), pp. 542-549, ISSN 0020-7136

Royal, I.; Lamarche-Vane, N.; Lamorte, L.; Kaibuchi, K. & Park, M. (2000). Activation of
cdc42, rac, PAK, and rho-kinase in response to hepatocyte growth factor
differentially regulates epithelial cell colony spreading and dissociation. *Molecular
Biology of the Cell*, Vol.11, No.5, (May 2000), pp. 1709-1725, ISSN 1059-1524

Rubin, J. S.; Bottaro, D. P. & Aaronson, S. A. (1993). Hepatocyte growth factor/scatter factor
and its receptor, the c-met proto-oncogene product. *Biochimica et Biophysica Acta*,
Vol.1155, No.3, (December 1993), pp. 357-371, ISSN 0006-3002

Saito, Y.; Nakagami, H.; Morishita, R.; Takami, Y.; Kikuchi, Y.; Hayashi, H.; Nishikawa, T.;
Tamai, K.; Azuma, N.; Sasajima, T. & Kaneda, Y. (2006). Transfection of human
hepatocyte growth factor gene ameliorates secondary lymphedema via promotion
of lymphangiogenesis. *Circulation*, Vol.114, No.11, (September 2006), pp. 1177-1184,
ISSN 0009-7322

Sakai, K.; Nakamura, T.; Matsumoto, K. & Nakamura, T. (2009). Angioinhibitory action of
NK4 involves impaired extracellular assembly of fibronectin mediated by perlecan-
NK4 association. *Journal of Biological Chemistry*, Vol.284, No.33, (August 2009), pp.
22491-22499, ISSN 0021-9258

Suda, T, & Liu, D. (2007). Hydrodynamic gene delivery: its principles and applications.
Molecular Therapy, Vol.15, No.12, (December 2007), pp. 2063-2069, ISSN 1525-0016

Tomioka, D.; Maehara, N.; Kuba, K.; Mizumoto, K.; Tanaka, M.; Matsumoto, K. &
Nakamura, T. (2001). Inhibition of growth, invasion, and metastasis of human
pancreatic carcinoma cells by NK4 in an orthotopic mouse model. *Cancer Research*,
Vol.61, No.20, (October 2001), pp. 7518-7524, ISSN 0008-5472

Underiner, T. L.; Herbertz, T. & Miknyoczki, S. J. (2010) Discovery of small molecule c-Met
inhibitors: Evolution and profiles of clinical candidates. *Anti-Cancer Agents in
Medicinal Chemistry*, Vol.10, No.1, (January 2010), pp. 7-27, ISSN 1871-5206

Wader, K.; Fagerli, U.; Holt, R.; Børset, M.; Sundan, A. & Waage, A. (2011). Soluble c-Met in
serum of multiple myeloma patients: correlation with clinical parameters. *European
Journal of Haematology*, in press, ISSN 0902-4441

Wang, R.; Zhang, J.; Chen, S.; Lu, M.; Luo, X.; Yao, S.; Liu, S.; Qin, Y. & Chen, H. (2011)
Tumor-associated macrophages provide a suitable microenvironment for non-small
lung cancer invasion and progression. *Lung Cancer*, in press, ISSN 0169-5002

Watabe, M.; Matsumoto, K.; Nakamura, T. & Takeichi, M. (1993). Effect of hepatocyte growth factor on cadherin-mediated cell-cell adhesion. *Cell Structure and Function,* Vol.18, No.2, (April 1993), pp. 117-1124, ISSN1347-3700

Wislez, M.; Rabbe, N.; Marchal, J.; Milleron, B.; Crestani, B.; Mayaud, C.; Antoine, M.; Soler, P. & Cadranel, J. (2003). Hepatocyte growth factor production by neutrophils infiltrating bronchioloalveolar subtype pulmonary adenocarcinoma: role in tumor progression and death. *Cancer Research,* Vol.63, No.6, (March 2003), pp. 1405-1412, ISSN 0008-5472

Wu, C. W.; Chi, C. W.; Su, T. L.; Liu, T. Y.; Lui, W. Y, & P'eng, F. K. (1998). Serum hepatocyte growth factor level associate with gastric cancer progression. *Anticancer Research,* Vol.18, No.5B, (September-Octorber 1998), pp. 3657-3659, ISSN 0250-7005

Weidner, K. M.; Arakaki, N.; Hartmann, G.; Vandekerckhove, J.; Weingart, S.; Rieder, H.; et al. (1991). Evidence for the identity of human scatter factor and human hepatocyte growth factor. *Proceeding of National Academy Sciences,* Vol.88, No.16, (August 1991), pp. 7001-7005, ISSN 0027-8424

Wen, J.; Matsumoto, K.; Taniura, N.; Tomioka, D. & Nakamura, T. (2004). Hepatic gene expression of NK4, an HGF-antagonist/angiogenesis inhibitor, suppresses liver metastasis and invasive growth of colon cancer in mice. *Cancer Gene Therapy,* Vol.11, No.6, (June 2004), pp. 419-430, ISSN 0929-1903

Wen, J.; Matsumoto, K.; Taniura, N.; Tomioka, D. & Nakamura, T. (2007). Inhibition of colon cancer growth and metastasis by NK4 gene repetitive delivery in mice. *Biochemical and Biophysical Research Communications,* Vol.358, No.1, (June 2007), pp. 117-123, ISSN 0006-291X

Wright, T. G.; Singh, V. K.; Li, J. J.; Foley, J. H.; Miller, F.; Jia, Z. & Elliott, B. E. (2009). Increased production and secretion of HGF alpha-chain and an antagonistic HGF fragment in a human breast cancer progression model. *International Journal of Cancer,* Vol.125, No.5, (September 2009), pp. 1004-1015, ISSN 0020-7136

Yancopoulos, G. D.; Klagsbrun, M. & Folkman, J. (1998). Vasculogenesis, angiogenesis, and growth factors: ephrins enter the fray at the border. *Cell,* Vol.93, No.5, (May 1998), pp. 661-664, ISSN: 0092-8674

Yano, S.; Wang, W.; Li, Q.; Matsumoto, K.; Sakurama, H.; Nakamura, T.; Ogino, H.; Kakiuchi, S.; Hanibuchi, M.; Nishioka, Y.; Uehara, H.; Mitsudomi, T.; Yatabe, Y.; Nakamura, T. & Sone, S. (2008). Hepatocyte growth factor induces gefitinib resistance of lung adenocarcinoma with epidermal growth factor receptor-activating mutations. *Cancer Research,* Vol.68, No.22, (November 2008), pp. 9479-9487, ISSN 0008-5472

Yoshinaga, Y.; Matsuno, Y.; Fujita, S.; Nakamura, T.; Kikuchi, M.; Shimosato, Y. & Hirohashi, S. (1993). Immunohistochemical detection of hepatocyte growth factor/scatter factor in human cancerous and inflammatory lesions of various organs. *Japanese Journal of Cancer Research,* Vol.84, No.11, (November 1993), pp. 1150-1158, ISSN 0910-5050

Zarnegar, R. & Michalopoulos, G. K. (1995). The many faces of hepatocyte growth factor: from hepatopoiesis to hematopoiesis. *The Journal of Cell Biology,* Vol.129, No.5, (June 1995), pp. 1177-1180, ISSN 0021-9525

Zeng, Q.; Chen, S.; You, Z.; Yang, F.; Carey, T. E.; Saims, D. & Wang, C. Y. (2002). Hepatocyte growth factor inhibits anoikis in head and neck squamous cell carcinoma cells by activation of ERK and Akt signaling independent of NFkappa B. *Journal of Biological Chemistry,* Vol.277, No.28, (July 2002), pp. 25203-25208, ISSN 0021-9258

Part 3

Detailed Experimental Analyses of Fluids and Flows

7

Hydrodynamics Influence on Particles Formation Using SAS Process

A. Montes, A. Tenorio, M. D. Gordillo,
C. Pereyra and E. J. Martinez de la Ossa
Department of Chemical Engineering and Food Technology,
Faculty of Science, UCA
Spain

1. Introduction

Particle size and particle size distribution play an important role in many fields such cosmetic, food, textile, explosives, sensor, catalysis and pharmaceutics among others. Many properties of industrial powdered products can be adjusted by changing the particle size and particle size distribution of the powder. The conventional methods to produce microparticles have several drawbacks: wide size distribution, high thermal and mechanical stress, environmental pollution, large quantities of residual organic solvent and multistage processes are some of them.

The application of supercritical fluids (SCF) as an alternative to the conventional precipitation processes has been an active field of research and innovation during the past two decades (Jung & Perrut, 2001; Martín& Cocero, 2008; Shariati &Peters, 2003).Through its impact on health care and prevention of diseases, the design of pharmaceutical preparations in nanoparticulate form has emerged as a new strategy for drug delivery. In this way, the technology of supercritical fluids allows developing micronized drugs and polymer-drug composites for controlled release applications; this also meets the pharmaceutical requirements for the absence of residual solvent, correct technological and biopharmaceutical properties and high quality (Benedetti et al., 1997; Elvassore et al., 2001; Falk& Randolph, 1998; Moneghini et al., 2001; Reverchon& Della Porta, 1999; Reverchon, 2002; Subramaniam et al., 1997; Yeo et al., 1993; Winters et al.,1996), as well as giving enhanced therapeutic action compared with traditional formulations (Giunchedi et al., 1998; Okada& Toguchi, 1995).

The revised literature demonstrates that there are two principal ways of micronizing and encapsulating drugs with polymers: using supercritical fluid as solvent, the RESS technique (Rapid Expansion of Supercritical Solutions); or using it as antisolvent, the SAS technique (Supercritical AntiSolvent); the choice of one or other depends on the high or low solubility, respectively, of the polymer and drug in the supercritical fluid.

Although the experimental parameters influences on the powder characteristic as particle size and morphologies is now qualitatively well known, the prediction of the powder characteristics is not feasible yet. This fact it is due to different physical phenomena involved in the SAS process. In most cases, the knowledge of the fluid phase equilibrium is necessary but not sufficient since for similar thermodynamic conditions, different hydrodynamics conditions can lead to different powder characteristics (Carretier et al., 2003).

So, the technical viability of the SAS process requires knowledge of the phase equilibrium existing into the system; the hydrodynamics: the disintegration regimes of the jet; the kinetics of the mass transfer between the dispersed and the continuous phase; and the mechanisms and kinetics of nucleation and crystal growth.

From the point of view of thermodynamics, the SAS process must satisfy the requirements outlined below. The solute must be soluble in an organic solvent but insoluble in the SCF. The solvent must also be completely miscible with the SCF, or two fluid phases would form and the solute would remain dissolved or partly dissolved in the liquid-rich phase. Thus, the SAS process exploits both the high power of supercritical fluids to dissolve organic solvents and the low solubility of pharmaceutical compounds in supercritical fluids to cause the precipitation of these materials once they are dissolved in an organic solvent, and thus spherical microparticles can be obtained.

On the other hand, characterization of hydrodynamics is relevant because of it is an important step for the success or the failure of the entire process, but with only some exception (Dukhin et al., 2005; Lora et al., 2000; Martín& Cocero, 2004), in the models developed for the SAS process, the hydrodynamics step received only limited consideration. For these reasons, the present review is focused on the investigation of the disintegration regime of the liquid jet into the supercritical (SC) CO_2. There are many works where correlations between the morphologies of the particles obtained in the drug precipitation assays and the estimated regimes were established (Carretier et al., 2003; Reverchon et al., 2010; Reverchon& De Marco, 2011; Tenorio et al., 2009). It was demonstrated that there are limiting hydrodynamic conditions that must be overcome to achieve a dispersion of the liquid solution in the dense medium; this dispersion must be sufficiently fine and homogeneous to direct the process toward the formation of uniform spherical nanoparticles and to the achievement of higher yields (Tenorio et al., 2009).

In this way, Reverchon et al. (Reverchon et al., 2010, Reverchon& De Marco, 2011) tried to find a correlation between particle morphology and the observed jet, concluding that expanded microparticles were obtained working at subcritical conditions; whereas spherical microparticles were obtained operating at supercritical conditions up to the pressure where the transition between multi- and single-phase mixing was observed. Nanoparticles were obtained operating far above the mixture critical pressure. However, the observed particle morphologies have been explained considering the interplay among high-pressure phase equilibria, fluid dynamics and mass transfer during the precipitation process, because in some cases the hydrodynamics alone is not able to explain the obtained morphologies, demonstrating the complexity of SAS processes. Moreover, the kinetics of nucleation and growth must also be considered.

2. Supercritical fluids

A supercritical fluid can be defined as a substance above its critical temperature and pressure. At this condition the fluid has unique properties, where it does not condense or evaporate to form a liquid or gas. A typical pressure-temperature phase diagram is shown in Figure 1. Properties of SCFs (solvent power and selectivity) can also be adjusted continuously by altering the experimental conditions (temperature and pressure). Moreover,

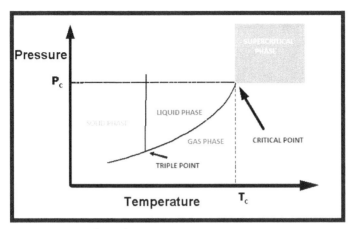

Fig. 1. Pressure-temperature phase diagram

these supercritical fluids have diffusivities that are two orders of magnitude larger than those of typical liquids, resulting in higher mass-transfer rates. Supercritical fluids show many exceptional characteristics, such as singularities in compressibility and viscosity, diminishing the differences between the vapor and liquid phases, and so on. Although a number of substances are useful as supercritical fluids, carbon dioxide has been the most widely used. Supercritical CO_2 avoids water discharge; it is low in cost, non-toxic and non-flammable. It has low critical parameters (304 K, 73.8 bar) and the carbon dioxide can also be recycled (Özcan et al., 1998).

3. Precipitation with SCF

The supercritical fluid technology has emerged as an important alternative to traditional processes of generation of micro and nanoparticles, offering opportunities and advantages such as higher product quality in terms of purity, more uniform dimensional characteristics, a variety of compounds to process and a substantial improvement on environmental considerations, among others.

Previously, it was discussed that the different particle formation processes using SCF are classified depending on how the SCF behaves, i.e., the supercritical CO_2 can play the role as antisolvent (AntiSolvent Supercritical process, SAS) or solvent (RESS process).

In the facilities of University of Cádiz, amoxicillin and ampicillin micronization have been carried out by SAS process (Montes et al., 2010, 2011a; Tenorio et al., 2007a, 2007b, 2008). Several experiments designs to evaluate the operating conditions influences on the particle size (PS) and particle size distribution (PSD) have been made. Pressures till 275 bar and temperatures till 338K have been used and antibiotic particle sizes have been reduced from 5-60 µm (raw material) to 200-500 nm (precipitated particles) (Figure 2).

The concentration was the factor that had the greatest influence on the PS and PSD. An increase in the initial concentration of the solution led to larger particles sizes with a wider distribution. Moreover, ethyl cellulose and amoxicillin co-precipitation has been carried out by SAS process (Montes et al., 2011b). SEM images of these microparticles are shown in Figure 3. It was noted that increasing temperature particle sizes were increased. Anyway, SEM images are not accurate enough to observe the distribution of both compounds

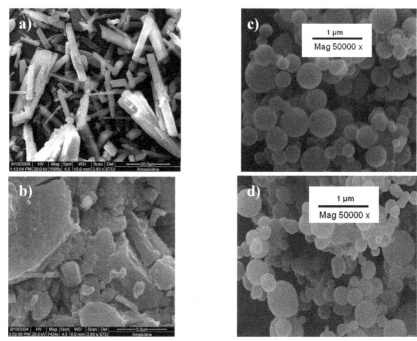

Fig. 2. SEM images of commercial a) amoxicillin and b) ampicillin, c) precipitated amoxicillin (Montes et al., 2010) and d) precipitated ampicillin (Montes et al., 2011a)

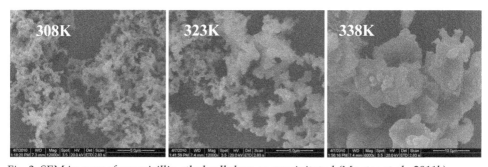

Fig. 3. SEM images of amoxicillin ethyl cellulose co-precipitated (Montes et al., 2011b).

because all the active substance could be situated on the surface of these microspheres and/or into the core. So, X-ray photoelectron spectroscopy (XPS) was used to determine the success of the encapsulation process by the chemical analysis of the particles on the precipitated surface (Morales et al., 2007). In this case, the elements that differentiate amoxicillin from ethyl cellulose are sulphur (S) and nitrogen (N) atoms. Therefore, these elements could indicate the location of the drug in the precipitated powders. On the other hand, amoxicillin delivery studies in simulated fluids from the co-precipitated obtained were carried out .The XPS spectra results were related to these drug delivery experiments and it was probed that the release of amoxicillin from precipitates in which N and S were present on the surface is faster than in cases these elements were not. Anyway, all the co-precipitated materials allowed a slower drug release rate than pure drug.

On the other hand, in the RESS method, the sudden expansion of supercritical solution (solute dissolved in supercritical carbon dioxide) via nozzle and the rapid phase change at the exit of the nozzle cause a high super-saturation, thus causing very rapid nucleation of the substrate in the form of very small particles that are collected from the gas stream. Hence, the conditions inside the expansion chamber are a key factor to control particle size and the particles grow inside the expansion chamber to their final size. This result clarifies the influence of two important process parameters on particle size. Both, a shorter residence time and, hence, less time available for particle growth as well as a higher dilution of the particles in the expansion chamber result in smaller particles.

3.1 Parameters influence on hydrodynamic

Mass transfer is one of the key factors that control the particle size in the SAS process. This is influenced by both the spray hydrodynamics of the organic solution and the thermodynamic properties of the supercritical fluid phase.

In the last years, the hydrodynamic of the SAS process has been the subject of several papers. Most authors face up to this problem considering that the jet of organic solvent behaves like a liquid jet injected into a gas, allowing to apply the classic theory of jet break-up. This theory could be applied successfully at subcritical conditions, below the mixture critical point solvent-CO_2, where there is surface tension. The mixture critical point denotes the limit of the two-phase region of the phase diagram. In other words, this is the point at which an infinitesimal change in some thermodynamic variable such as temperature or pressure will lead to separation of the mixture into two distinct phases.

However, in supercritical conditions, above the critical point of the mixture organic solvent and CO_2, it is not possible to distinguish droplets nor interfaces between the liquid solution and the phase of dense CO_2 gas. Surface tension decreases to zero in a shorter distance than characteristic break-up lengths. Thus, the jet spreads forming a gaseous plume and will be characterized by the degree of turbulence associated with the vortices produced in the SC CO_2 (Chehroudi et al., 2002; Kerst et al., 2000; Reverchon et al., 2010). Lengsfeld et al. were the first group that investigated fluid dynamics of the SAS process, studying the evolution and disappearance of the liquid surface tension of fluids injected in supercritical carbon dioxide. They concluded that a gas-like jet is formed after the jet break-up (Lengsfeld et al., 2000). In this way, Kerst et al. determined the boundaries between the different modes and they noted a strong interdependence between mass transfer and fluid dynamics (Kerst et al., 2000).

In the SAS related literature there is a general agreement about the flow regimes observable when a liquid is injected in a vessel. The way in which the liquid solution is dispersed in the CO_2 when the operating conditions are below the mixture critical point (MCP), which is strongly influenced by the operating pressure and the flow rate of liquid solution at fixed temperature, can be described according to one of the following four regimes: 1) the dripping mode, which requires lower flow speed so that drops can detach themselves from the orifice, 2) the Rayleigh break up regime, which is characterized by a rupture of the jet in the form of monodisperse droplets, 3) the sine wave break up regime, in which a helicoidal oscillation of the jet occurs, leading to its rupture into droplets with a polydisperse distribution, and 4) atomization, in which the jet is smooth when it leaves the orifice, until it reaches the zone of highly chaotic rupture where a cone of atomized liquid is formed.

When SAS is performed at supercritical conditions a transition between multi-phase and single-phase mixing is observed by increasing the operating pressure. Single-phase mixing is due to the very fast disappearance of the interfacial tension between the liquid solvent and the fluid phase in the precipitator. The transition between these two phenomena depends on the operating pressure, but also on the viscosity and the surface tension of the solvent. Reverchon et al. demonstrates that in the case of dimethyl sulfoxide (DMSO) at pressures larger than the MCP a progressive transition exists between multi-phase and single-phase mixing, but is not observed, even for pressures very close to the MCP, in the case of acetone (Reverchon et al., 2010). In the dripping mode, the droplet size decrease with increase in pressure operation due to a corresponding decrease in the interface tension, so the initial droplet size can be manipulated by small changes in the pressure of CO_2 (Lee et al., 2008).

However, in the Rayleigh disintegration mode, the droplet size is weakly dependent on the interface tension of the system and is proportional to the diameter of the jet. In the dripping mode, the size and shape of the drops become highly dependent on the nozzle exit condition.

Sometimes, the transition between multi-phase (formation of droplets after jet break-up) and single-phase mixing (no formation of droplets after jet break-up) could not be located at the pressure of the mixture critical point. Dukhin et al. (Dukhin et al., 2003) and Gokhale et al. (Gokhale et al., 2007) found that jet break-up into droplets still takes place at pressures slightly above the MCP. Due to the non-equilibrium conditions during mixing, there is a dynamic (transient) interfacial tension that decreases between the inlet of the liquid and its transformation to a gas-like mixture. The transition between these multi-phase and single-phase mixing depends on the operating pressure, but also on the viscosity and the surface tension of the solvent.

Not only the thermodynamics but also the nozzle device or liquid solution flow rate will influence on the observed regime. The kind of injection device and its orifices diameter will determine the chosen liquid solution flow rate to get a successful jet break up. In this way, in a previous work, when the 200 μm diameter nozzle was used with a liquid flow rate of 1mL/min, the solution was not atomized, and we did not obtain any precipitation (Tenorio et al., 2009).

A lot of parameters control the precipitation process and many particle morphologies have been observed. As it was commented before, the kind of injection device used (and its efficiency), can strongly influence the precipitation process. The objective of these devices in SAS processing is to produce a very large contact surface between the liquid and the fluid phase, to favour the mass transfer between the antisolvent and the liquid solvent inducing jet break-up and atomization of the liquid phase.

Various injection devices to produce liquid jet break-up have been proposed in the literature. Yeo et al. (Yeo et al., 1993) proposed the adoption of a nozzle and tested various nozzle diameters ranging from 5 to 50 μm. Moussa et al. (Moussa et al., 2005) showed that the pressure distribution during the expansion of the supercritical fluid is a function of the nozzle length and diameter. Other authors used small internal diameter capillaries (Dixon et al., 1993; Randolph et al., 1993). Coaxial devices have also been proposed: in the SEDS process (solution enhanced dispersion by supercritical fluids) a coaxial twin-fluid nozzle to co-introduce the SCF antisolvent and solution is used (Bałdyga et al., 2010; He et al., 2010; Mawson et al., 1997; Wena et al., 2010). Complex nozzles geometries have also been tested carrying out a comparative study of the nozzle by computational fluid dynamics (Balabel et al., 2011; Bouchard et al., 2008). Petit-Gas et al. found that for the lowest capillary internal diameter studied, there were particles with differences morphologies according to the jet velocity. For the lowest jet velocity, irregular morphology was obtained, and for highest jet velocity spherical morphology was obtained (Petit-Gas et al., 2009). However, for the highest capillary internal diameter experiments, particles morphology difference was less important. Particles were quasi-spherical, to a lesser extent for the smallest jet velocity. Once more time it was demonstrated the parameters interrelation in SAS process and its great complexity. Not only the kind of nozzle but also the nozzle relative position to CO_2 inlet must be taken into account. In this way, Martin & Cocero studied the differences on hydrodynamics and mixing when CO_2 is not introduced through the concentric annulus, but through a different nozzle, which is placed relatively far from the nozzle of the organic solution. Since the inlet velocity of CO_2 is much lower than the inlet velocity of the solution, this flow has a relatively small influence on hydrodynamics and mixing. However, if CO_2 is not introduced through the annulus, the fluid that diffuses into the jet is no longer almost pure CO_2, but fluid from the bulk fluid phase, which has some amount of organic solvent. This greatly reduces the supersaturation and bigger particles are formed (Martin & Cocero, 2004).

Moreover, these different unstable modes (Rayleigh break up, sine wave break up and atomization) are controlled by several competing effects: capillary, inertial, viscous, gravity and aerodynamic effects (Petit-Gas et al., 2009). The predominance of each effect has been discussed in several works (Badens et al., 2005; Carretier et al., 2003; Kerst et al., 2000). Reynolds number gives a measure of the ratio of inertial forces to viscous forces. For the lower Reynolds numbers, Rayleigh regime is observed and surface tension is the chief force controlling the break-up of an axisymmetrical jet. For higher Reynolds numbers, the inertial forces compete with the capillary forces. There is a lateral motion in the jet break-up zone which leads to the formation of an asymmetrical jet, which can be either sinuous or helicoidal. Finally, when the flow rate goes beyond a certain value, the aerodynamic effects become quite strong and the jet is atomised. Another dimensionless number frequently used to describe jet fluid dynamics is the Ohnesorge (Oh) number that relates the viscous and the surface tension force by dividing the square root of Weber number by Reynolds number (Badens et al., 2005; Czerwonatis, 2001; Kerst et al., 2000).

In this way, taking into account the critical atomization velocity defined as the velocity corresponding to the boundary between the asymmetrical mode and the atomization mode, it is possible to tune the process towards one or another regime. Moreover this critical velocity seems to be dependent on CO_2 density. Badens et al. observed a decrease in this critical jet velocity when the CO_2 continuous phase density increases (Badens et al., 2005). Badens et al. and Czerwonatis et al. found out the predominant effect of the continuous phase properties on jet break-up, especially in the asymmetrical and direct atomization modes because of the aerodynamic forces preponderance (Badens et al., 2005; Czerwonatis et al., 2001). However Petit-Gas et al. concluded that variations of the continuous phase properties had no effects on the transition velocity in the studied conditions (Petit-Gas et al., 2009).

3.2 Morphology

Some authors attempted to connect the observed flow or mixing regimes to the morphology of the precipitated particles. Lee et al. injected a solution of dichloromethane (DCM) and poly lactic acid (PLA) at subcritical conditions in the dripping and in the Rayleigh disintegration regimes and observed the formation of uniform PLA microparticles (Lee et al., 2008). Other authors (Chang et al., 2008; Gokhale et al., 2007; Obrzut et al., 2007; Reverchon et al., 2008) did not find relevant differences in the various precipitates obtained. Particularly, PLA morphologies showed to be insensitive to the SAS processing conditions (Randolph et al., 1993). This characteristic fact could be assigned to the high molecular weights and the tendency to form aggregated particles because of the reduction of the glass transition temperature in SC-CO_2.

At subcritical conditions the interfacial tension between the injected liquid and the bulk phase never goes to zero and a supercritical mixture is not formed between the liquid solvent and CO_2. The droplets formed during atomization are subjected to a very fast internal formation of a liquid/CO_2 mixture. Due to a high solubility of CO_2 in pressurized organic liquids and a very poor evaporation of organic solvents into the bulk CO_2, the droplets expand. During these processes, the interfacial tension allows the droplets to maintain its spherical shape, even when the solute is precipitated within the droplet. Saturation occurs at the droplet surface and solidification takes place with all solutes progressively condensing on the particle internal surface. The final result is the formation of a solid shell.

This kind of particles has also been observed in other SAS works (Reverchon et al., 2008). It has been also obtained expanded hollow particle at same conditions. The different surface morphologies can depend on different controlling mass transfer mechanisms, as suggested by Duhkin et al. (Duhkin et al., 2005).

Operating conditions above the MCP, from a thermodynamic point of view, are characterized by zero interfacial tension. But, the liquid injected into the precipitator, before equilibrium conditions are obtained, experiences the transition from a pure liquid to a supercritical mixture. Therefore, interfacial tension starts from the value typical of the pure liquid and progressively reduces to zero. This fact means that droplets formed after jet break-up (whose presence indicates in every case the existence of an interfacial tension) are formed before the disappearance of the interfacial tension. In other words, the time of equilibration is longer than the time of jet break-up and spherical microparticles instead of nanoparticles can be obtained.

3.3 Visualization techniques

Many researchers have used imaging and visualization techniques to study jet flows, atomization, and droplets; a number of systems are reviewed in the literature (Bell et al., 2005; Chigier et al.,1991). Jet lengths and spray widths ranging to milimeters and drop and particle sizes ranging to micrometers must be taking into account in order to select imaging system components.

Several studies used particle and droplet visualization in supercritical fluids (Badens et al., 2005; Gokhale et al.,2007; Kerst et al.,2000; Lee et al.,2008; Mayer & Tamura,1996; Obrzut et al.,2007; Randolph, et al., 1993; Shekunov et al., 2001).

The optical technique described in these works provides the ability to visualize mixing occurring between two fluids with different refractive indices. For instance, shadowgraphy is an optical method to obtain information on non-uniformities in transparent media, independently if they arise by temperature, density or concentration gradients. All of these inhomogeneities refract light which causes shadows.

Although for SAS precipitation, microscopy-base imaging offers the advantage of examining the dynamic process that leads to particle formation, the presence of particles smaller than two microns complicates an already difficult task of imaging an injection process.

The ability to identify and characterize these small formations drives future system improvements, including lighting enhancements laser-induced fluorescence, and higher spatial resolution cameras. In this way Reverchon et al. used light scattering technique to clearly differentiate between an atomized very droplet laden spray and a dense "gas-plume", limitation which cannot be gained by applying optical techniques due to the fact that both the droplet laden spray and the dense "gas-plume" result in a dark shadow (Reverchon et al., 2010).

On the other hand, extensive research has been done using scanning electron microscopy (SEM) to evaluate the size and morphology of particles formed under supercritical conditions (Armellini& Tester, 1994; Bleich et al., 1994; Mawson et al. 1997; Randolph et al., 1993; Shekunov et al., 2001;). A limitation of SEM analysis is that it is applied to particles after they have been removed from the dynamic system.

4. A particular case: Ampicillin SAS precipitation

In our research group a study was carried out to establish a correlation between the morphologies of the particles obtained in the ampicillin precipitation assays and the estimated regimes. This correlation would be an ideal tool to establish the limiting hydrodynamic conditions for the success of the test in order to define the successful experiments; that is, the appropriate conditions to orientate the process toward the formation of uniform spherical nanoparticles instead of irregular and larger-sized particles, for the solute-solvent-SC CO_2 system studied (Tenorio et al.,2009).

A series of ampicillin precipitation experiments by the SAS technique, utilizing N-methyl-pyrrolidone (NMP) as the solvent and CO_2 as the antisolvent, under different operating conditions were carried out. Two nebulizers, with orifice diameters of 100 and 200 µm, respectively were used.

A pilot plant, developed by Thar Technologies® (model SAS 200) was used to carry out all the experiments. A schematic diagram of this plant is shown in Figure 4. The SAS 200 system comprises the following components: two high-pressure pumps, one for the CO_2 (P1) and the other for the solution (P2), which incorporate a low-dead-volume head and check valves to provide efficient pumping of CO_2 and many solvents; a stainless steel precipitator vessel (V1) with a 2L volume consisting of two parts, the main body and the frit, all surrounded by an electrical heating jacket (V1-HJ1); an automated back-pressure regulator (ABPR1) of high precision, attached to a motor controller with a position indicator; and a jacketed (CS1-HJ1) stainless steel cyclone separator (CS1) with 0.5L volume, to separate the solvent and CO_2 once the pressure was released by the manual back-pressure regulator (MBPR1).The following auxiliary elements were also necessary: a low pressure heat exchanger (HE1), cooling lines, and a cooling bath (CWB1) to keep the CO_2 inlet pump cold and to chill the pump heads; an electric high-pressure heat exchanger (HE2) to preheat the CO_2 in the precipitator vessel to the required temperature quickly; safety devices (rupture discs and safety valve MV2); pressure gauges for measuring the pump outlet pressure (P1, PG1), the precipitator vessel pressure (V1, PG1), and the cyclone separator pressure (CS1, PG1); thermocouples placed inside (V1-TS2) and outside (V1-TS1) the precipitator vessel, inside the cyclone separator (CS1-TS1), and on the electric high pressure heat exchanger to obtain continuous temperature measurements; and a FlexCOR Coriolis mass flowmeter (FM1) to measure the CO_2 mass flow rate and another parameters such as total mass, density, temperature, volumetric flow rate, and total volume.

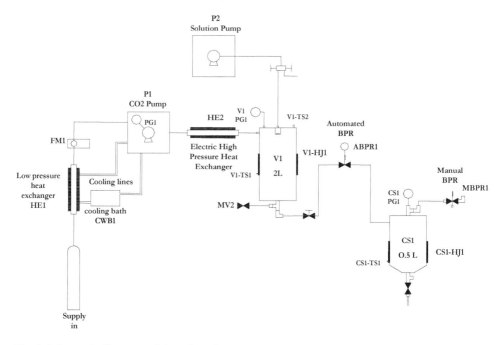

Fig. 4. Schematic diagram of the pilot plant

The pendant droplet method, as introduced by Andreas and Tucker, was used to determine the interfacial tension between NMP and SC CO_2 (Andreas&Tucker, 1938).This method, and its application to high pressures and temperatures, are comprehensively described by Jaeger (Jaeger et al., 1996). A commercial CCD video technique allows recording of droplet shapes for subsequent video image processing.

Rayleigh breakup, sinusoidal wave break up, and atomization regimes are seen to be clearly differentiated by representing graphically the Reynolds number against Ohnesorge number Here, the forces of inertia of the liquid phase (pressure gradient), the forces of capillarity (surface tension), and those of viscosity of the liquid phase (friction) are taken into account, but the force of gravity is considered to be negligible.

Two differentiated types of morphology can be identified in the precipitated experiments: spherical nanoparticles of ampicillin that are obtained from a fine precipitate with foamy texture, and particles of ampicillin with irregular forms and larger size, which are characteristic of the precipitate formed by aggregates, compact films, and rods (Figure 5).

The aim of the work is to explain, from the estimation of the different disintegration regimes as a function of the physicochemical properties and of the velocity of the jet, the two different morphologies obtained in the ampicillin precipitation experiments for a specific range of operating conditions. Thus it should be possible to specify the hydrodynamic conditions for orientating the process toward the formation of uniform spherical nanoparticles rather than larger size irregular particles.

The morphology of the precipitate obtained at low pressure was supposed to be in accordance with the Rayleigh estimated regime, since droplets with a diameter of approximately twice the diameter of the orifice would be produced; (Badens et al., 2005)

| 80 bar | 100 bar | 125 bar | 150 bar |

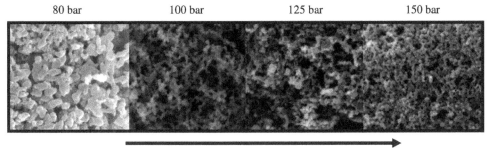

Increasing Operating Pressure

Fig. 5. Effect of operating pressure on microstructure of ampicillin powder obtained by the SAS experiments (Tenorio et al., 2009).

then, because sufficient contact area would not be generated, the liquid phase does not evaporate in the dense phase of the CO_2. Instead, the liquid droplets accumulate in the filter, where the precipitate is obtained by the drying action of the CO_2.

In contrast, for higher pressures, the presence of a precipitate occurring as aggregates in the filter may be explained by the existence of significant mechanisms that stabilize the liquid jet. These important mechanisms of stabilization may be associated with the existence of the dynamic interfacial tension (Dukhin et al., 2003).Therefore, the so-called "gaseous plume" or "gas-like jet", which is characteristic of states of complete miscibility of mixtures (above their MCP),would not be produced, even at 150 bar.

The influence of the mean velocity of the jet of liquid solution was also analyzed. The liquid solution flow rate from 1 mL/min to 5 mL/min causes the jet to disintegrate, passing through the three possible regimes: Rayleigh, sine wave break-up and atomization. The lowest flow rate tested (1 mL/min), which is equivalent to a jet velocity of 0.5 m/s (200 µm nozzle diameter), led to an unsatisfactory test result, which may be in agreement with the Rayleigh-type estimated regime; this is because the droplets that formed would not generate sufficient contact area to produce saturation while they are in motion, and, consequently, ampicillin is not precipitated. When the liquid solution flow rate is increased to 2 mL/min a dispersion of the sine wave breakup type is estimated. Considering that a polydisperse

Fig. 6. SEM images showing the microstructure of the ampicillin powder obtained by SAS experiment with 5ml/min (at 180 bar, 328 K, and 200 µm) (Tenorio et al., 2009).

distribution of droplets is produced in this regime, it is very well correlated with the experimental obtained results (Tenorio et al., 2009).

When the flow rate is increased to 3 mL/min, it is estimated that the transition is complete, and the liquid is atomized. The large quantity of fine precipitate with foamy texture obtained both on the walls and accumulated in the filter (characteristic of nanoparticles) would have originated from the fully atomized and homogeneous dispersion that is occurring in the precipitation chamber. With 5 mL/min it was obtained similar results in accordance with the estimated atomization regime (Figure 6).

5. Conclusions

The hydrodynamics of the SAS process has been revised. Nozzle device, liquid flow rate and pressure effects on hydrodynamics have been taken into account. Flow regimes observable in the SAS related literature have been described. Dripping mode is simply due to the use of liquid flow rates that are too low to produce a continuous liquid flow and do not produce atomization. Rayleigh breakup, sinusoidal wave break up, and atomization regimes and, particularly their competition at some process conditions require a detailed analysis. The ability to identify and characterize these regimes drives future system improvements, including lighting enhancements laser-induced fluorescence, and higher spatial resolution cameras.

Morphology of the precipitated particles can be related to flow or mixing regimes. In the ampicillin case, two differentiated types of morphology can be identified in the precipitated experiments: spherical nanoparticles of ampicillin that are obtained from a fine precipitate with foamy texture, and particles of ampicillin with irregular forms and larger size, which are characteristic of the precipitate formed by aggregates, compact films, and rods. It has been correlated the morphologies of the particles obtained in the ampicillin precipitation assays and the estimated regimes as a function of the physicochemical properties and of the velocity of the jet, for a specific range of operating conditions.

However, the results from the application of these correlations cannot explain the morphologies of the precipitates obtained in some experiments. This fact can be due to important stabilization mechanisms as dynamic interfacial tension

Due to the great complexity of the SAS process, factors such as the ternary phase equilibrium, matter transfer between the phases, and the kinetics of nucleation and growth need to be considered, in addition to the limiting hydrodynamic conditions.

6. Acknowledgment

We are grateful to the Spanish Ministry of Education and Science (Project No. CTQ2010-19368) for financial support.

7. References

Andreas, J. M.; Tucker, W. B. (1938). Boundary Tension by Pendant Drop.*J. Phys. Chem.*,42, pp. 1001-1019.

Armellini, F.J., Tester, J.W. (1994). Precipitation of sodium chloride and sodium sulfate in water from sub- to supercritical conditions:150 to 550°C, 100 to 300 bar. *J. Supercrit. Fluids*, 7, pp.147–158.

Badens, E., Boutin, O., Charbit, G. (2005). Laminar jet dispersion and jet atomization in pressurized carbon dioxide, *J. Supercrit. Fluids* ,36, pp. 81–90.

Balabel, A., Hegab, A.M., Nasr, M., El-Behery, S. M. (2011). Assessment of turbulence modeling for gas flow in two-dimensional convergent–divergent rocket nozzle. *Appl. Math. Model.*, 35, pp. 3408–3422.

Bałdyga, J., Kubicki, D., Shekunov, B.Y., Smithd, K.B. (2010). Mixing effects on particle formation in supercritical fluids, *Chem. Eng. Res. Des.*, 88, pp. 1131–1141.

Bell, P. W., Stephens, A. P.,Roberts, C. B., Duke, S. R. (2005). High-resolution imaging of the supercritical antisolvent process, *Experiments in Fluids*, 38, pp. 708–719.

Benedetti, L., Bertucco, A., Pallado, P. (1997). Production of micronic particles of biocompatible polymer using supercritical carbon dioxide, *Biotechnol. Bioeng.*, 53, pp. 232-237.

Bleich, J., Cleinebudde, P., Muller, B.W. (1994). Influence of gas density and pressure on microparticles produced with the ASES process, *Int. J. Pharm.*, 106, pp. 77–84.

Bouchard, A., Jovanovic, N., A. H. de Boer, Martín, A., Jiskoot, W., Crommelin, D. J.A. , Hofland, G.W., Witkamp, G.-J. (2008). Effect of the spraying conditions and nozzle design on the shape and size distribution of particles obtained with supercritical fluid drying, *Eur. J. Pharm. Biopharm.*, 70, pp. 389–401.

Carretier, E., Badens, E., Guichardon, P., Boutin, O., Charbit, G. (2003). Hydrodynamics of supercritical antisolvent precipitation: characterization and influence on particle morphology, *Ind. Eng. Chem. Res.*, 42, pp. 331–338.

Czerwonatis, N., Eggers, R., Charbit, G. (2001). Disintegration of liquid jets and drop drag coefficients in pressurized nitrogen and carbon dioxide, *Chem. Eng. Tech.*, 24, pp. 619–624.

Chang, S.-C., Lee, M.-J., Lin, H.-M. (2008). Role of phase behavior in micronization of lysozyme via a supercritical anti-solvent process, *Chem. Eng. J.*, 139, pp. 416–425.

Chehroudi, B., Cohn, R., Talley, D. (2002). Cryogenic shear layers: experiments and phenomenological modelling of the initial growth rate under subcritical and supercritical conditions, *Int. J. Heat Fluid Flow*, 23, pp. 554-563.

Chigier, N. (1991). Optical imaging of sprays. *Prog. Energy Combust. Sci.*, 17, pp.211–262

Dixon, D.J., Johnston, K.P., Bodmeier, R.A. (1993).Polymeric materials formed by precipitation with a compressed fluid antisolvent, *AIChE J.*, 39 (1), pp. 127–139.

Dukhin, S.S., Zhu, C., Pfeffer, R., Luo, J.J., Chavez, F., Shen, Y. (2003). Dynamic interfacial tension near critical point of a solvent–antisolvent mixture and laminar jet stabilization, *Physicochem. Eng. Aspects*, 229, pp. 181-199.

Dukhin, S. S., Shen, Y., Dave, R., Pfeffer, R. (2005). Droplet mass transfer, intradroplet nucleation and submicron particle production in two-phase flow of solvent-supercritical antisolvent emulsion. *Colloids Surf. A*, 261, pp. 163-176.

Elvassore, N., Bertucco, A., Caliceti, P. (2001). Production of insulin-loaded poly(ethylene glycol)/poly(l-lactide) (PEG/PLA) nanoparticles by gas antisolvent techniques, *J. Pharm. Sci.*, 90, pp. 1628-1636.

Falk, R.F., Randolph, T.W. (1998). Process variable implications for residual solvent removal and polymer morphology in the formation of gentamycin-loaded poly (L-lactide) microparticles, *Pharm. Res.*, 15, pp. 1233-1237.

Giunchedi, P., Genta, I., Conti, B., Conte, U., Muzzarelli, R.A.A. (1998) .Preparation and characterization of ampicillin loaded methylpyrrolidinone chitosan and chitosan microspheres, *Biomat.*, 19, pp.157-161.

Gokhale, A., Khusid, B., Dave, R.N., Pfeffer, R. (2007).Effect of solvent strength and operating pressure on the formation of submicrometer polymer particles in supercritical microjets, *J. Supercrit. Fluids* ,43, pp 341-356.

He, W., Jiang, Z., Suo, Q.,Li, G. (2010). Mechanism of dispersing an active component into a polymeric carrier by the SEDS-PA process, *J. Mater. Sci.*, 45, pp. 467–474.

Jaeger, Ph T.; Schnitzler, J. V.; Eggers, R. (1996). Interfacial Tension of Fluid Systems Considering the Non-Stationary Case with Respect to Mass Transfer. *Chem. Eng. Technol.*, 19, pp.197.

Jung, J., Perrut., M. (2001). Particle design using supercritical fluids: Literature and patentsurvey. *J. Supercrit. Fluids*, 20 (3), pp. 179-219.

Kerst, A.W., Judat, B., Schlünder, E.U. (2000).Flow regimes of free jets and falling films at high ambient pressure, *Chem. Eng. Sci.*, 55, pp. 4189-4208.

Lee, L.Y., Lim L. K., Hua, J., Wang, C.-H. (2008). Jet breakup and droplet formation in near-critical regime of carbon dioxide-dichloromethane system, *Chem.Eng. Sci.*, 63, pp. 3366-3378.

Lengsfeld, C.S., Delplanque, J.P., Barocas, V.H., Randolph, T.W. (2000).Mechanism governing microparticle morphology during precipitation by a compressed antisolvent: atomization vs nucleation and growth, *J. Phys. Chem. B*, 104, pp. 2725–2735.

Lora, M., Bertucco, A., Kikic, I. (2000). Simulation of the Semicontinuous Supercritical Antisolvent Recrystallization Process. *Ind. Eng. Chem. Res.*, 39, pp. 1487-1496.

Martín, A., Cocero, M.J. (2004). Numerical modeling of jet hydrodynamics, mass transfer, and crystallization kinetics in the supercritical antisolvent (SAS) process *J. Supercrit. Fluids*, 32, pp. 203–219.

Martín, A., Cocero, M.J. (2008).Precipitation processes with supercritical fluids: patents review, *Recent Patents Eng.* 2, pp.9 -20.

Mawson S, Kanakia S, Johnston KP. (1997). Coaxial nozzle for control of particle morphology in precipitation with a compressed fluid antisolvent, *J. Appl. Polym. Sci.*, 64:, pp. 2105-2118.

Mayer, W.,Tamura, H. (1996). Propellant injection in a liquid oxygen/ gaseous hydrogen rocket engine. *J. Propul. Power*, 12(6), pp.1137–1147.

Moneghini, M., Kikic I., Voinovich, D., Perissutti, B., Filipovic-Grcic, J. (2001). Processing of carbamazepine - PEG 4000 solid dispersions with supercritical carbon dioxide: Preparation, characterisation, and in vitro dissolution, *Int. J. Pharm.*, 222, pp.129-138.

Montes, A., Tenorio, A., Gordillo, M.D., Pereyra, C., Martínez de la Ossa, E. (2010). Screening design of experiment applied to supercritical antisolvent precipitation of amoxicillin: exploring new miscible conditions. *J. Supercrit. Fluids*, 51, pp. 399-403.

Montes, A., Tenorio, A., Gordillo, M.D., Pereyra, C., Martínez de la Ossa, E. (2011a). Supercritical Antisolvent Precipitation of Ampicillin in Complete Miscibility Conditions. *Ind. Eng. Chem. Res.*, 50, pp. 2343-2347.

Montes, A., Gordillo, M.D., Pereyra, C., Martínez de la Ossa, E. (2011b). Co-precipitation of amoxicillin and ethylcellulose microparticles by supercritical antisolvent process *J. Supercrit. Fluids* (in press).

Morales, M. E., Ruiz, M. A., Oliva, I., Oliva, M., Gallardo, V. (2007). Chemical characterization with XPS of the surface of polymer microparticles loaded with morphine. *Int. J. Pharm.*, 333, pp. 162-166.

Moussa, A. B., Ksibi, H., Tenaud, C., Baccar, M. (2005). Parametric study on the nozzle geometry to control the supercritical fluid expansion, *Int. J. Thermal Sciences*, 44, pp. 774-786.

Obrzut, D.L., Bell, P.W., Roberts, C.B., Duke, S.R. (2007). Effect of process conditions on the spray characteristics of a PLA plus methylene chloride solution in the supercritical antisolvent precipitation process, *J. Supercrit. Fluids*, 42, pp. 299-309.

Okada, H., Toguchi, H. (1995). Biodegradable microspheres in drug delivery, *Crit. Rev. in Ther. Drug Carrier Syst.*, 12, pp. 1-99.

Ozcan, A.S.; Clifford, A.A.; Bartle, K.D. and Lewis, D.M. (1998). Dyeing of cotton fibres with disperse dyes in supercritical carbon dioxide, *Dyes and Pigments*, 36(2), pp. 103-110.

Petit-Gas, T., Boutin, O., Raspo, I., Badens, E. (2009). Role of hydrodynamics in supercritical antisolvent processes, *J. Supercrit. Fluids* , 51, pp. 248-255.

Randolph, T.W., Randolph, A.D., Mebes, M., Yeung, S. (1993). Sub-micrometer-sized biodegradable particles of poly(L- lactic acid) via the gas antisolvent spray precipitation process, *Biotechnol. Progress*, 9, pp. 429-435.

Reverchon, E., Della Porta, G. (1999). Production of antibiotic micro- and nano-particles by supercritical antisolvent precipitation, *Powder Technol.*, 106, pp. 23-29.

Reverchon, E. (2002). Supercritical-assisted atomization to produce micro- and/or nanoparticles of controlled size and distribution, *Ind. Eng. Chem. Res.*, 41, pp. 2405-2411.

Reverchon, E., Adami, R., Caputo, G., De Marco, I. (2008).Expanded microparticles by supercritical antisolvent precipitation: interpretation of results. *J. Supercrit.Fluids*, 44(1), pp. 98-108.

Reverchon, E.; Torino, E.; Dowy, S.; Braeuer, A.; Leipertz, A. (2010). Interactions of phase equilibria, jet dynamics and mass transfer during supercritical antisolvent micronization. *Chem. Eng. J.*, 156, pp. 446-458.

Reverchon, E., De Marco, I. (2011). Mechanisms controlling supercritical antisolvent precipitate morphology. *Chem. Eng. J.* (In press).

Shariati, A., Peters. C. J. (2003). Recent developments in particle design using supercritical fluids, *Curr. Opin. Solid State Mater. Sci.*, 7, (4–5) pp. 371-383.

Shekunov, B. Yu., Baldyga, J., York, P. (2001). Particle formation by mixing with supercritical antisolvent at high Reynolds numbers. *Chem. Eng. Sci.*, 56, pp. 2421-2433.

Subramaniam, B., Snavely, K., Rajewski, R.A. (1997). Pharmaceutical processing with supercritical carbon dioxide, J. Pharm. Sci., 86, pp. 885-890.

Tenorio, A., Gordillo, M. D., Pereyra, C.M., Martínez de la Ossa, E.J. (2007a). Controlled submicro particle formation of ampicillin by supercritical antisolvent precipitation, *J. Supercrit. Fluids*, 40, pp. 308-316.

Tenorio, A.; Gordillo, M. D.; Pereyra, C. M.; Martínez de la Ossa, E. M. (2007b). Relative importance of the operating conditions involved in the formation of nanoparticles of ampicillin by supercritical antisolvent precipitation. *Ind. Eng. Chem. Res.*, 46, pp. 114-123.

Tenorio, A., Gordillo, M. D., Pereyra, C.M., Martínez de la Ossa, E.J. (2008). Screening design of experiment applied to supercritical antisolvent precipitation of amoxicillin, *J. Supercrit. Fluids*, 44, pp. 230-237.

Tenorio, A., Jaeger, P., Gordillo, M.D., Pereyra, C.M., Martínez de la Ossa, E.J. (2009). On the selection of limiting hydrodynamic conditions for the SAS process, *Ind. Eng. Chem. Res.*,48 (20), pp. 9224-9232.

Wena, Z., Liua, B., Zhenga, Z., Youa, X., Pua, Y., Li, Q. (2010). Preparation of liposomes entrapping essential oil from Atractylodes macrocephala Koidz by modified RESS technique, *Chem. Eng. Res. Des.*, 88, pp. 1102–1107.

Winters, M.A., Knutson, B.L., Debenedetti, P.G., Sparks, H.G., Przybycien, T.M., Stevenson, C.L., Prestrelski, S.J. (1996). Precipitation of proteins in supercritical carbon dioxide, *J. Pharm. Sci.*, 85, pp. 586-594.

Yeo, S.-D., Lim, G.-B., Debenedetti, P.G., Bernstein, H. (1993). Formation of microparticulate protein powders using a supercritical fluid antisolvent, *Biotechnol. Bioeng.*, 41, pp. 341-346.

8

Flow Instabilities in Mechanically Agitated Stirred Vessels

Chiara Galletti and Elisabetta Brunazzi
Department of Chemical Engineering,
Industrial Chemistry and Materials Science, University of Pisa
Italy

1. Introduction

A detailed knowledge of the hydrodynamics of stirred vessels may help improving the design of these devices, which is particularly important because stirred vessels are among the most widely used equipment in the process industry.

In the last two decades there was a change of perspective concerning stirred vessels. Previous studies were focused on the derivation of correlations able to provide global performance indicators (e.g. impeller flow number, power number and mixing time) depending on geometric and operational parameters. But recently the attention has been focused on the detailed characterization of the flow field and turbulence inside stirred vessels (Galletti et al., 2004a), as only such knowledge is thought to improve strongly the optimization of stirred vessel design.

The hydrodynamics of stirred vessels has resulted to be strongly three dimensional, and characterised by different temporal and spatial scales which are important for the mixing at different levels, i.e. micro-mixing and macro-mixing.

According to Tatterson (1991) the hydrodynamics of a mechanically agitated vessel can be divided at least into three flow systems:

- *impeller flows* including discharge flows, trailing vortices behind the blades, etc.;
- *wall flows* including impinging jets generated from the impeller, boundary layers, shed vortices generated from the baffles, etc.;
- *bulk tank flows* such as large recirculation zones.

Trailing vortices originating behind the impeller blades have been extensively studied for a large variety of impellers. For instance for a Rushton turbine (RT) they appear as a pair, behind the lower and the upper sides of the impeller blade, and provide a source of turbulence that can improve mixing. Assirelli et al. (2005) have shown how micro-mixing efficiency can be enhanced when a feeding pipe stationary with the impeller is used to release the fed reactant in the region of maximum dissipation rate behind the trailing vortices. Such trailing vortices may also play a crucial role in determining gas accumulation behind impeller blades in gas-liquid applications, thus affecting pumping and power dissipation capacity of the impeller.

But in the last decade lots of investigations have pointed out that there are other important vortices affecting the hydrodynamics of stirred vessels. In particular it was found that the flow inside stirred vessels is not steady but characterised by different flow instabilities, which can influence the flow motion in different manners. Their knowledge and comprehension is still far from complete, however the mixing optimisation and safe operation of the stirred vessel should take into account such flow variations.

The present chapter aims at summarizing and discussing flow instabilities in mechanically agitated stirred vessels trying to highlight findings from our research as well as from other relevant works in literature. The topic is extremely wide as flow instabilities have been detected with different investigation techniques (both experimental and numerical) and analysis tools, in different stirred vessel/impeller configurations.

Thus investigation techniques and related analysis for the flow instability detection will be firstly overviewed. Then a possible classification of flow instabilities will be proposed and relevant studies in literature will be discussed. Finally, examples of findings on different flow instabilities and their effects on the mixing process will be shown.

2. Investigation techniques

Researchers have employed a large variety of investigation techniques for the detection of flow instabilities. As such techniques should allow identifying flow instabilities, they should be able to detect a change of the flow field (or other relevant variables) with time. Moreover a good time resolution is required to allow an accurate signal processing. Regarding this point, actually flow instabilities in stirred vessels are generally low frequencies phenomena as their frequency is much smaller than the impeller rotational frequency N; so, effectively, the needed temporal resolution is not so high. Anyway the acquisition frequency should at least fulfil the Nyquist criterion.

The graph of Fig. 1a summarises the main techniques, classified as experimental and numerical, employed so far for the investigation of flow instabilities. A brief description of the techniques will be given in the following text in order to highlight the peculiarities of their applications to stirred vessels.

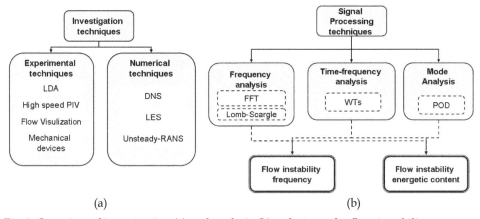

(a)   (b)

Fig. 1. Overview of investigation (a) and analysis (b) techniques for flow instability characterization in stirred vessels.

2.1 Experimental techniques

Laser Doppler anemometry (LDA) is one of the mostly used experimental technique for flow instability detection. LDA is an optical non-intrusive technique for the measurement of the fluid velocity. It is based on the Doppler shift of the light scattered from a 'seeding' particle, which is chosen to be nearly neutrally buoyant and to efficiently scatter light. LDA does not need any calibration and resolves unambiguously the direction of the velocity. Moreover it provides high spatial and temporal resolutions. These are very important for flow instability detection. In addition, more than one laser Doppler anemometer can be combined to perform multi-component measurements. The application of LDA to cylindrical stirred vessels requires some arrangement in order to minimize refraction effects at the tank walls, so often the cylindrical vessel is placed inside a square trough.

Particle Image Velocimetry (PIV) is also an optical technique which allows the velocity of a fluid to be simultaneously measured throughout a region illuminated by a two-dimensional light sheet, thus enabling the instantaneous measurements of two velocity components. However recently the use of a stereoscopic approach allows all three velocity components to be recorded. So far the temporal resolution of PIV measurements has been limited because the update rate of velocity measurements, governed by the camera frame rate and the laser pulse rate, was too low. Thus PIV was not suited for the investigation of flow instabilities. However recently, high-frame rate PIV systems have been developed allowing flow measurements with very high update rates (more than 10 kHz); thus its use for the analysis of flow instabilities in stirred vessels has been explored by some investigators. Similarly to LDA, also PIV requires the fluid and vessel walls to be transparent as well as actions to minimize refraction effects at the tank curvature.

Different flow visualization techniques have also been used to help clarifying the mechanism of flow instabilities. Such flow visualization techniques may simply consist of tracing the fluid with particles and recording with a camera a region of the flow illuminated by a laser sheet. More sophisticated techniques are able of providing also concentration distribution: for example Laser Induced Fluorescence, LIF, uses a fluorescent marker and a camera (equipped with a filter corresponding to the wave of fluorescence) which detects the fluorescence levels in the liquid.

In addition to such optical instruments, different mechanical devices have been used in literature for the detection of flow instabilities. Such devices are based on the measurements of the effect of flow instabilities on some variables. Bruha et al. (1995) employed a "tornadometer", that is a device which allows measuring the temporal variation of the force acting on a small target placed into the flow where instabilities are thought to occur. Paglianti et al. (2006) proved that flow instabilities in stirred vessels could be detected by pressure transducers positioned at the tank walls. The pressure transducers provided time series of pressure with a temporal resolution suited for the flow instability detection. Such a technique is particularly interesting as it is well suited for industrial applications. Haam et al. (1992) identified flow instabilities from the measurement of heat flux and temperature at the walls through heat flux sensors and thermocouples. Hasal et al. (2004) measured the tangential force acting on the baffles as a function of time by means of mechanical devices. Also power number measurements (as for instance through strain gauge techniques) have been found to give an indication of flow instabilities related to change in the circulation loop (Distelhoff et al., 1995).

2.2 Numerical techniques

Numerical models have also been used for the investigation of flow instabilities in stirred vessels, especially because of the increasing role of Computational Fluid Dynamics (CFD). Logically, since the not steady nature of such instabilities, transient calculation techniques have to be employed. These may be classified in: Unsteady Reynolds-averaged Navier-Stokes equations (URANS), Large Eddy Simulation (LES) and Direct Numerical Simulation (DNS)

URANS employs the usual Reynolds decomposition, leading to the Reynolds-averaged Navier-Stokes equations, but with the transient (unsteady) term retained. Subsequently the dependent variables are not only a function of the space coordinates, but also a function of time. Moreover, part of the turbulence is modelled and part resolved. URANS have been applied to study stirred vessels by Torré et al. (2007) who found indications on the presence of flow instabilities from their computations; however their approach was not able to identify precessional flow instabilities.

LES consists of a filtering operation, so that the Navier-Stokes are averaged over the part of the energy spectrum which is not computed, that is over the smaller scales. Since the remaining large-scale turbulent fluctuations are directly resolved, LES is well suited for capturing flow instabilities in stirred vessels, although it is very computationally expensive. This has been shown for both single-phase (for example Roussinova et al., 2003, Hartmann et al., 2004, Nurtono et al. 2009) and multi-phase (Hartmann et al., 2006) flows.

DNS consists on the full resolution of the turbulent flow field. The technique has been applied by Lavezzo et al. (2009) to an unbaffled stirred vessel with Re = 1686 providing evidence of flow instabilities.

3. Analysis techniques

The above experimental or modelling investigations have to be analysed with suited tools in order to get information on flow instabilities. These consist mainly of signal processing techniques, which are applied to raw data, such as LDA recordings of the instantaneous velocity, in order to gain information on the characteristics of flow instabilities.

Two kinds of information have been extracted so far:
-   frequency of the flow instabilities as often they appear as periodic phenomena;
-   relevance of flow instabilities on the flow motion.

Among the techniques which have been employed in literature for the characterization of flow instabilities in stirred vessels, there are (see Fig. 1b):
-   frequency analysis techniques (the Fast Fourier Transforms and the Lomb-Scargle periodogram method);
-   time-frequency analysis techniques (Wavelet Transforms);
-   principal component analysis (Proper Orthogonal Decomposition).

Whereas the first two techniques have been largely used for the determination of the flow instability frequency, the latter has been used to evaluate the impact of flow instabilities on the motion through the analysis of the most energetic modes of the flow.

3.1 Frequency analysis

The Fast Fourier Transform (FFT) decomposes a signal in the time domain into sines and cosines, i.e. complex exponentials, in order to evaluate its frequency content. Specifically the FFT was developed by Cooley & Tukey (1965) to calculate the Fourier Transform of a K samples series with $O(Klog_2K)$ operations. Thus FFT is a powerful tool with low computational demand, but it can be performed only over data evenly distributed in time. In case of LDA recordings, these should be resampled and the original raw time series replaced with series uniform in time. As for the resampling techniques, simple methods like the "Nearest Neighbour" or the "Sample and Hold" should be preferred over complex methods (e.g. "Linear Interpolation", "Spline Interpolation"), because the latter bias the variance of the signal. It should be noticed that the resampled series contains complete information about the spectral components up to the Nyquist critical frequency $fc=1/2\Delta$ where Δ in the sampling interval. At frequencies larger than the Nyquist frequency the information on the spectral components is aliased.

The Lomb-Scargle Periodogram (LSP) method (Lomb, 1976, Scargle, 1982) performs directly on unevenly sampled data. It allows analysing frequency components larger than the Nyquist critical frequency: this is possible because in irregularly spaced series there are a few data spaced much closer than the average sampling rate, removing ambiguity from any aliasing. The method is much more computational expensive than FFTs, requiring $O(10^2 K^2)$ operations.

It is worthwhile discussing the suitability of the analysis techniques described above for the investigation of flow instabilities and what are the main parameters to be considered. Flow instabilities are low frequency phenomena, therefore we are interested in the low frequency region of the frequency spectrum. The lowest frequency which can be resolved with both the FFT and Lomb-Scargle method is inversely related to the acquisition time; hence longer sampling times yield better frequency resolutions. This explains the long observations made for flow instabilities detection in stirred vessels. In our works on flow instabilities we have used typically LDA recordings at least 800 s long. In other words the sampling time should be long enough to cover a few flow instabilities cycles. As the time span covered by a series is proportional to the number of samples, the application of the LSP to long series requires strong computational effort.

A benchmark between the two methods is provided in Galletti (2005) and shown in Fig. 2.

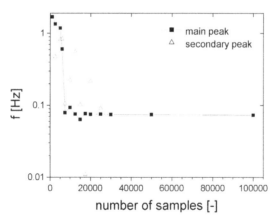

Fig. 2. Frequency of the main and the secondary peak in the low frequency region of the spectrum calculated with the Lomb-Scargle method as a function of the number of samples. RT, D/T = 0.33, C/T = 0.5, Re = 27,000. Galletti (2005).

The solid squares show the frequency f of the main peak identified in the spectrum calculated with the LSP as function of the number of samples used for the analysis. It can be observed that f is scattered for low numbers of samples, and it approaches asymptotically the value of f = 0.073 Hz (the same of the FFT analysis over the whole acquisition time of 800 s with 644,000 samples) as the number of samples increases. The empty triangles indicate the presence of further low frequency peaks. The main fact to be aware of is that low time intervals conceal the flow instabilities by covering only a portion of the fluctuations.

3.2 Time-frequency analysis

Both FFT and LSP inform how much of each frequency component exists in the signal, but they do not tell us when in time these frequencies occur in the signal. For transient flows it may be of interest the time localisation of the spectral component. The Wavelet Transform (WT) is capable of providing the time and frequency information simultaneously, hence it gives a time-frequency representation of the signal (Daubechies, 1990, Torrence and Compo, 1998). The WT breaks the signal into its "Wavelets", that are functions obtained from the scaling and the shifting of the "mother Wavelet" ψ. The WT has been proposed for the investigation of stirred vessels by Galletti et al. (2003) and subsequently applied by Roy et al. (2010).

3.3 Proper orthogonal decomposition

POD is a linear procedure, based on temporal and spatial correlation analysis, which allows to decompose a set of signals into a modal base, with the first mode being the most energetic (related to large-scale structures thus trailing vortices and flow instabilities) and the last being the least energetic (smaller scales of turbulence). It was first applied for MI characterisation by Hasal et al. (2004) and latterly by Ducci & Yianneskis (2007). An in-depth explanation of the methodology is given in Berkooz et al. (1993).

4. Classification of flow instabilities

A possible classification of flow instabilities in stirred vessels is reported in Fig. 3. The graph is not aimed at imposing a classification of flow instabilities, however it suggests a way of interpretation which may be regarded as a first effort to comprehend all possible instabilities.

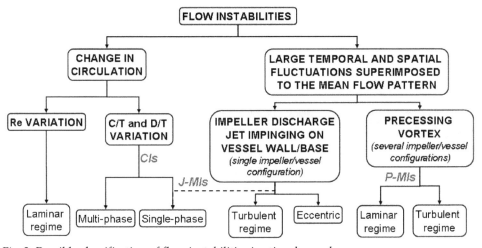

Fig. 3. Possible classification of flow instabilities in stirred vessels.

4.1 Change in circulation pattern

A first kind of flow instability (see left-hand side of the diagram of Fig. 3) manifests as a real change in the circulation pattern inside the tank. Two main sources of such a change have been identified: a variation of the Reynolds number (Re) or a variation of the impeller/vessel geometrical configuration.

In relation to the former source, Nouri & Whitelaw (1990) reported a transition due to Re variations in the flow pattern induced by a 60° PBT with D = T/3 set at C = T/3 in a vessel of T = 0.144 m. For non-Newtonian fluids a flow pattern transition from a radial to an axial flow was observed as the Re was increased up to Re = 4,800. For Newtonian fluids the authors observed that the flow pattern transition occurred at about Re = 650. This value was also confirmed by the power number measurements through strain-gauges carried out by Distelhoff et al. (1995). Similar investigations on such transition may be found in the works of Hockey (1990) and Hockey & Nouri (1996).

Schäfer et al. (1998) observed by means of flow visualisation the flow discharged by a 45° PBT to be directed axially at higher Re and radially at lower Re. The flow stream direction was unstable, varying from radial to axial, for Re = 490-510. A similar flow transition was also indicated by Bakker et al. (1997) who predicted with CFD techniques the flow pattern generated from a 4-bladed 45° PBT of diameter D = T/3 and set at C = T/3 inside a tank of T = 0.3 m. The regime was laminar, the Reynolds number being varied between 40 and 1,200. The impeller discharge stream was directed radially for low Re numbers, however for Re larger than 400 the flow became more axial, impinging on the vessel base rather than on the walls.

A second source of instabilities, manifesting as a flow pattern change, is associated with variations of the impeller/vessel geometrical configuration, which means either variations of the distance of the impeller from the vessel bottom (C/T) or variation of the impeller diameter (D/T) or a combination of both variations.

This kind of instabilities were firstly noticed by Nienow (1968) who observed a dependency on the clearance of the impeller rotational speed required to suspend the particles (Njs) in a solid-liquid vessel equipped with a D = 0.35T RT. He observed that for C < T/6 the pattern was different (the discharge stream was directed downwards towards the vessel corners) from the typical radial flow pattern, providing low Njs values. Baldi et al. (1978) also observed a decrease of the Njs with the impeller off-bottom clearance for a 8-bladed turbine. Conti et al. (1981) found a sudden decrease of the power consumption associated with the change in the circulation pattern when lowering the impeller clearance of a 8-bladed turbine. The aforementioned authors concluded that the equation given by Zwietering (1958) for the calculation of the Njs should be corrected in order to take into account the dependency on C/T.

The dependency of the power number on the impeller off-bottom clearance was also observed by Tiljander & Theliander (1993), who measured the power consumption of two PBT of different sizes, i.e. D = T/3 and D = T/2, and a high flow impeller of D = T/2. The visual observation of the flow pattern revealed that at the transition point between the axial and the radial flow patterns, the circulation inside the vessel appears chaotic.

Ibrahim & Nienow (1996) investigated the efficiency of different impellers, i.e. a RT, a PBT pumping either upwards or downwards, a Chemineer HE3 and a Lightnin A310 hydrofoil pumping downwards and a Ekato Intermig agitator, for solids suspension. For the RT, the aforementioned authors observed a sudden decrease of both the impeller speed and the mean dissipation rate required to just suspend the particles as the clearance was decreased from C = T/3 down to C = T/6 for the impeller having D = T/3; such a clearance corresponded to the transition from the radial flow pattern to the axial.

Subsequently, a strong influence of the clearance on the suspension of particles was confirmed also by Myers et al. (1996) for three axial impellers. If the clearance was sufficiently high the discharge flow impinged on the vessel wall rather then the base, leading to a secondary circulation loop which was directed radially inward at the vessel base and returned upwards to the impeller at the centre of the vessel. Such a reversed flow occurred for C > 0.45T for a PBT of diameter D = 0.41T and for C > 0.25T for a straight-blade turbine of the same diameter, whereas only for very high clearances (C > 0.95T) for a high efficiency Chemineer impeller having the same diameter.

Bakker et al. (1998) reported that the flow pattern generated by either a PBT or a three-blade high efficiency impeller depended on C/T and D/T, influencing the suspension of the particles.

Armenante & Nagamine (1998) determined the Njs and the power consumption of four impellers set at low off-bottom clearances, typically C < T/4. For radial impellers, i.e. a RT and a flat blade turbine, they observed that the clearance at which the change in the flow pattern from a radial to an axial type occurred was a function of both impeller type and size, i.e. D/T. In particular the flow pattern changed at lower C/T for larger impellers. This was in contrast with previous works (see for example Conti et al., 1981) which reported a clearance of transition independent on D/T. For instance Armenante & Nagamine (1998) found the flow pattern transition to occur at 0.16 < C/T < 0.19 for a Rushton turbine with a diameter D = 0.217T and at 0.13 < C/T < 0.16 for a D = 0.348T RT. For the flat blade turbine the clearances at which the transition took places were higher, being of 0.22-0.24 and 0.19-0.21 for the two impeller sizes D = 0.217T and D = 0.348T, respectively.

Sharma & Shaikh (2003) provided measurements of both Njs and power consumption of solids suspension in stirred tanks equipped with 45° PBT with 4 and 6 blades. They plotted the critical speed of suspension Njs as a function of C/T distinguishing three regions, according to the manner the critical suspension speed varied with the distance of the impeller from the vessel base. As the impellers were operating very close to the vessel base, the Njs was observed to be constant with C/T (first region); then for higher clearances Njs increased with C/T because the energy available for suspension decreased when increasing the distance of the impeller from the vessel base (second region), and finally (third region) for high clearances the Njs increased with C/T with a slope higher than that of the second region. The onset of third region corresponded to the clearance at which the flow pattern changed from the axial to the radial flow type. In addition the aforementioned authors observed that as the flow pattern changed the particles were alternatively collected at the tank base in broad streaks and then suddenly dispersed with a certain periodicity. They concluded that a kind of instabilities occurred and speculated that maybe the PBT behaved successively as a radial and axial flow impellers.

The influence of C on the flow pattern has been intensively studied also for single-phase flow in stirred tanks. Yianneskis et al. (1987) showed that the impeller off-bottom clearance affects the inclination of the impeller stream of a Rushton turbine of diameter D = T/3. In particular the discharge angle varied from 7.5° with respect to the horizontal plane for C = T/4 down to 2.5° for C = T/2.

Jaworski et al. (1991) measured with LDA the flow patterns of a 6-bladed 45° PBT having a diameter D = T/3 for two impeller clearances, C = T/4 and C = T/2. For the lower impeller clearance, the impeller stream impinged on the vessel base and generated an intensive radial circulation from the vessel axis towards the walls. For the higher clearance the impeller stream turned upwards before reaching the base of the vessel, generating also a reverse flow directed radially from the walls towards the vessel axis at the base of the vessel.

Kresta & Wood (1993) measured the mean flow field of a vessel stirred with a 4-bladed 45° PBT for two impeller sizes, i.e. D = T/3 and D = T/2, and varying the impeller clearance systematically from T/20 up to T/2. They observed that the circulation pattern underwent a transition at C/D = 0.6, and for the larger impeller (D = T/2) such a transition was accompanied by a deflection of the inclination of the discharge stream toward the horizontal.

Ibrahim & Nienow (1995) measured the power number of different impellers for a wide range of Reynolds number, i.e. 40 < Re < 50,000, in Newtonian fluids. For a D = T/3 RT they observed that the power numbers with clearances of C = T/3 and C = T/4 was the same for all Re; however for C = T/6 the discharge flow was axial rather than radial and the associated power number was considerably lower (by about 25%) for all the range of Re investigated. For a D = T/2 RT a radial discharge flow was still observed at C = T/6 for all Re except for those with the highest viscosity (1 Pa·s).

Rutherford et al. (1996a) investigated the flow pattern generated by a dual Rushton impeller and observed different circulation patterns depending on the impeller clearance of the lower impeller and the separations between the two impellers, observing three stable flow patterns: "parallel flow", "merging flow" and "diverging flow" patterns.

Mao et al. (1997) measured with LDA the flow pattern generated from various PBT of different sizes in the range of 0.32 < D/T < 0.6 and number of blades varying from 2 to 6 in a stirred vessel in turbulent regime (Re > 20,000). They used two impeller off-bottom clearances, C = T/3 and C = T/2, observing a secondary circulation loop with the higher clearance.

Montante et al. (1999) provided a detailed investigation of the flow field generated by D = T/3 RT placed at different off-bottom clearances varying from C = 0.12T to C = 0.33T. They found that the conventional radial flow pattern (termed "double-loop" pattern) occurred for C = 0.20T, but it was replaced by an axial flow pattern (termed "single loop" pattern) as the clearance was decreased to C = 0.15T. A reduction of the power number from 4.80-4.85 for C/T = 0.25-0.33 down to 3.80 as the clearance was decreased to C/T = 0.12-0.15 was reported, so that the power consumption was reduced by about 30% as the flow underwent a transition from the double- to the single-loop pattern.

4.1.1 Clearance instabilities (CIs)

Gallettl et al. (2003, 2005a, 2005b) studied the flow pattern transition for a D = T/3 RT and identified a kind of flow instabilities, which will be denoted as CIs (clearance instability). The authors found that the flow pattern transition (single- to double-loop pattern) occurred for C/T = 0.17-0.2, thus within an interval of clearances of about 0.03T. Such C interval was dependent on the fluid properties, lower clearances being observed for more viscous fluids. At clearances of flow pattern transition the velocity time series indicated flow pattern instabilities as periods of double-loop regime, single-loop regime and "transitional" state that followed each other. When the flow underwent a change from one type of circulation to another, the transitional state was always present and separated in time the single- from the double-loop flow regime. Nevertheless, a flow pattern could change firstly into the transitional state and afterwards revert to the original flow regime, without changing the type of circulation. The occurrence of the three flow regimes was shown to be random, and their lifetimes could be significant, often of the order of few minutes. The time duration of the three flow regimes depended on the impeller clearance, higher clearances promoting the double-loop regime. Moreover the time duration of the three flow regimes depended on the impeller rotational speed, higher impeller rotational speeds promoting the double-loop regime.

An example of flow pattern transition is shown in the LDA time series of Fig. 4a which indicated different regimes, that can be attributed to the double-loop, transitional and single-loop patterns. The most surprising finding was that within the transitional state an instability was manifested as a periodic fluctuation of the flow between the double and the single-loop regimes, characterised by a well-defined frequency f. Such frequency was linearly dependent on the impeller speed according to $f' = f/N = 0.12$.

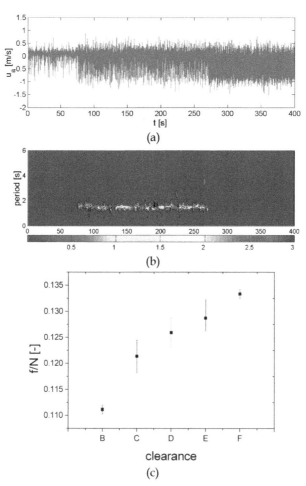

Fig. 4. Wavelet power analysis of axial velocity data: (a) time series; (b) Wavelet power spectrum; (c) dependence of frequency on impeller clearance (B is the highest, F the lowest clearance). Taken from Galletti et al. (2003).

Therefore the flow pattern transition which occurs for a RT when changing the impeller position is governed by two types of instability. The first one manifests as a random succession of double-loop regime, single-loop regime and transitional state over large time intervals. The second one is the instability encountered during the transitional state, characterised by a well-defined periodicity of the order of few seconds.

The exact nature of the clearance-related instabilities is not fully understood, but it is not likely to be related to the turbulence content of the flows, as the phenomenon is characterised by a single frequency even for the lowest Re range studied with the most viscous fluid, for which Re is around 5,200 and the corresponding flows should be mostly laminar. Some evidences as the increase of f' with lowering C/T (or increasing the impeller stream mean velocity by reducing the impeller blade thickness to diameter ratio t_b/D) may confirm that it is the interaction between the impeller discharge stream and the vessel base/walls to play a major role in the formation of such instability.

4.2 Macro-instabilities

Another kind of instability (see the right-hand side of the diagram of Fig. 3) manifests itself as large temporal and spatial variations of the flow superimposed to the mean flow patterns, thus such flow instabilities are called "macro-instabilities". On the basis of results achieved during our work and from other works in literature it was chosen to divide this kind of flow instability into two subgroups, because we think that there were two different underlying mechanisms driving such instabilities.

4.2.1 Precessional macro-instabilities (P-MIs)

The first subgroup comprehends flow instabilities which seem to be associated with a vortex moving about the shaft. The first evidence of this vortex was provided by Yianneskis et al. (1987) who noticed that the vortex motion produced large temporal and spatial fluctuations superimposed on the mean flow pattern. A similar vortex was also observed by Haam et al. (1992) cited earlier.

Precessional MIs were investigated by Nikiforaki et al. (2003), who used two different impellers (RT and PBT) having the same diameter D = T/3 for Re > 20,000. The frequency of the macro-instabilities was found to be linearly related to the impeller speed with f' = f/N = 0.015-0.020, independently on impeller clearance and design. In a more recent work Nikiforaki et al. (2004) studied the effect of operating parameters on macro-instabilities. In particular they observed the presence of other frequencies varying from f'= 0.04-0.15 , as the Reynolds number was reduced.

Hartmann et al. (2004) performed a LES simulation of the turbulent flow (Re = 20,000 and 30,000) in a vessel agitated with a D = T/3 RT set at C = T/2. The geometries of the vessel and impeller were identical to those used for the experiments of Nikiforaki et al. (2003). The simulation indicated the presence of a vortical structure moving round the vessel centreline in the same direction as the impeller. Such structure was observed both below and above the impeller (axial locations of z/T = 0.12 and z/T = 0.88 were monitored), however the two vortices were moving with a mutual phase difference. The frequency associated with the vortices was calculated to be f' = 0.0255, therefore slightly higher than the 0.015-0.02 reported by Nikiforaki et al. (2003). The authors concluded that this may encourage an improvement of the sub-scale grid and/or the numerical settings.

Importantly, the presence of a phase shift between the precessing vortices below and above the impeller was confirmed by the LDA experiments of Micheletti & Yianneskis (2004).

These authors used a cross-correlation method between data taken in the upper and lower parts of the vessel, and estimated a phase difference between the vortices in the two locations of approximately 180°.

The presence of the precessing vortex was assessed also in a solid-liquid system by the LES simulation of Derksen (2003).

Hasal et al. (2004) investigated flow instabilities with a Rushton turbine and a pitched blade turbine, both of D = T/3 with the proper orthogonal decomposition analysis. They confirmed the presence of the precessing vortex, however they found different f' values depending on the Re. In particular f' values akin to those of Nikiforaki et al. (2003) were observed for high Re, whereas higher values, i.e. f' = 0.06-0.09 were found for low Re.

Galletti et al. (2004b) investigated macro-instabilities stemming from the precessional motion of a vortex about the shaft for different impellers, geometries and flow regimes. The authors confirmed that the P-MI frequency is linearly dependent on the impeller rotational speed, however they indicated that different values of the proportionality constant between MI frequency and impeller rotational speed were found for the laminar and turbulent flow regimes, indicating different behaviour of MIs depending on the flow Re (see Fig. 5a). For intermediate (transitional) regions two characteristic frequencies were observed, confirming the presence of two phenomena. In particular in the laminar flow region P-MIs occurred with a non-dimensional frequency f' about 7-8 times greater than that observed for the turbulent region. This was proved for two RTs (D/T = 0.33 and 0.41 RT) as well as for a D/T = 0.46 PBT. Thus the impeller design does not affect P-MIs for both laminar and turbulent regions. The impeller off-bottom clearance does not affect significantly the P-MI frequency for the Rushton turbine and the pitched blade turbine (see for instance Fig. 5b). However differences in the regions where P-MIs are stronger may be found, as for instance lower impeller clearances originated weaker P-MIs near the liquid surface.

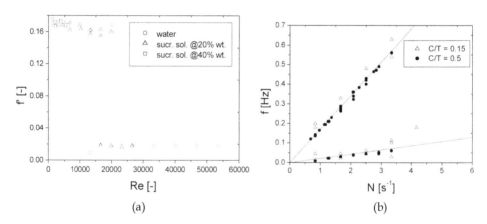

(a)   (b)

Fig. 5. (a) Non-dimensional macro-instability frequency as a function of the impeller Reynolds number. RT, D/T = 0.41, C/T = 0.5. (b) Macro-instability frequency as a function of the impeller rotational speed for different clearances. RT, D/T = 0.41. Galletti (2005).

Importantly, Galletti et al. (2004b) found that the MI frequency is affected by the impeller diameter. For the laminar regime a linear dependence of the non-dimensional macro-instability frequency on the impeller to tank diameter ratio was established:

$$f' = a \cdot \left(\frac{D}{T}\right) + b \tag{1}$$

A deep clarification of precessional MIs triggering mechanism in both laminar and turbulent regimes was provided by Ducci & Yianneskis (2007) for a D = T/3 RT placed at C = T/2. The authors used 2-point LDA and a 2-D PIV with a 13kHz camera. Through a vortex identification and tracking technique, the authors showed that P-MIs stem from a precessional vortex moving around the vessel axis with f' = 0.0174 for the turbulent regime. In laminar regime the frequency corresponding to a precession period was higher, of about f' = 0.13. The slight differences on the frequencies with the work of Galletti et al. (2004b) may be imputed to the different spectral analysis. For instance in the vortex tracking method the frequency was evaluated from the time needed to a vortex to complete 360°, whereas the FFT analysis of Galletti et al. (2004b) covered several MI cycles. But importantly Ducci & Yianneskis (2007) showed that in the laminar regime the vortex precessional motion was much closer to the axis than in turbulent regime (for which the vortex tends to stay rather far from the axis). In addition to that the authors showed a change in the flow pattern between the laminar and turbulent conditions, which affects the precessional MI frequency.

In a later work Ducci et al. (2008) investigated also the transitional regime showing the interaction between the two frequency instabilities (f' = 0.1 and f' = 0.02 of the laminar and turbulent regime, respectively). They found that the two simultaneous instabilities are associated to two different types of perturbation of the main mean flow: an off-centering instability that results in a precession of the vortex core centre with a f' = 0.02 and a stretching instability that induces an elongation of the vortex core along a direction that is rotating with f' = 0.1 around the vessel axis. For higher Re, the authors identified an interaction between the perturbations of the mean vortex core associated to f' = 0.02 off-centering structures and a f' = 0.04 stretching/squeezing instability.

A deep investigation of precessional MIs was also carried out by the same group (Doulgerakis et al., 2011) for an axial impeller (PBT) with D = T/2 placed at C = T/2 with Re = 28,000. The MI frequency distribution across the vessel indicated the presence of many frequencies reported before in literature. However the two dominant frequencies were f' = 0.1 and f' = 0.2. The POD analysis showed that the first mode can be seen as a radial off-center perturbation of the mean flow that results in a precession of the vortex core around the impeller axis with f' = 0.1. The second mode is an instability which stretches/squeezes the vortex core in a direction that is rotating with f' = 0.1. Importantly also for the PBT, the higher frequency was exactly double than the lower one as for the RT case. This would be also in agreement with many spectral analysis reported in Galletti (2005) which showed the presence of an additional peak frequency about the double of the P-MI frequency.

Kilander et al. (2006) identified through LSP analysis of LDA data frequencies with f' = 0.025 for the turbulent regime (thus in fully agreement with the work by Hartmaan et al., 2006) in a vessel agitated by a D = T/3 RT.

Lately, many other computational methods confirmed also the presence of precessional MIs. Nurtono et al. (2009) obtained from LES simulations a frequency f' = 0.0125 for a D=T/3 RT placed at C = T/2 for Re = 40,000.

The DNS simulations of Lavezzo et al. (2009) for an unbaffled vessel equipped with a 8-blade paddle impeller indicated the presence of a spiralling vortex with f' = 0.162 for Re = 1686. The application of Eq. [1] developed by Galletti et al. (2004b) to the above case would give a higher frequency $f' = 0.24$, however it should be pointed out that the equation was developed for baffled configurations.

4.2.2 Jet impingement macro-instabilities (J-MIs)

Other evidence of large temporal and spatial variations of the flow macro-instabilities have been reported in the last decade and they not always seem to be related to a precessional vortex.

Bruha et al. (1995) used a device called "tornadometer" to estimate the flow instabilities induced by a 6-bladed 45° PBT of D = 0.3T set at C = 0.35T. The target was axially located above the impeller at $z/C = 1.2$ and 1.4 and at radial distance equal to the impeller radius. The aforementioned authors found a linear relation between the instability frequency f and the impeller rotational speed N, according to f = -0.040 N +0.50. In a later work (Bruha et al., 1996) the same authors reported a linear dependence of the MI frequency on N ($f' = 0.043$-0.0048) for Re values above 5,000. No flow-instabilities were noted for Re < 200 and an increase in f' was observed for 200 < Re < 5,000.

Montes et al. (1997) studied with LDA the flow instabilities in the vicinity of the impeller. induced by a 6-bladed 45° PBT of D = 0.33T set at C = 0.35T and observed different values for f' depending on the Reynolds number: $f' = 0.09$ for Re = 1140 and $f' = 0.0575$ for Re = 75,000. They suggested that macro-instabilities appear as the switching between one loop and two or many loops, taking place between the impeller and the free surface and they are able to alter this surface. This leads to different flow patterns in front of the baffles or between two adjacent baffles. The mechanism is complex and three-dimensional but the large vortices clearly appear in a regular way, with a well defined frequency. Hasal et al. (2000) used the proper orthogonal decomposition to analyse LDA data observed for a PBT and found a $f' = 0.087$ for Re = 750 and Re = 1,200, and a value of 0.057 for Re = 75,000. In addition they noticed that the fraction of the total kinetic energy carried by the flow instabilities (relative magnitude) varied with the location inside the stirred vessel, they being stronger in the central and wall regions below the impeller but weaker in the discharge flow from the impeller.

Myers et al. (1997) used digital PIV to investigate flow instabilities in a stirred tank equipped with two different impellers: a 4-bladed 45° PBT of D = 0.35T and a Chemineer HE-3 of D = 0.39T. The PBT was set at C = 0.46T and 0.33T, whereas the Chemineer HE-3 was set at C = 0.33T. The Reynolds number was ranging between 6,190 and 13,100. For the higher clearance, i.e. C = 0.46T, the PBT showed flow fluctuations of about 40 s for an impeller rotational speed N = 60 rpm, therefore $f' = 0.025$. The same impeller set at the lower clearance, C = 0.33T, showed more stable flow fields, with not very clear peaks in the low frequency region of the spectra, at around $f' = 0.07$-0.011. The Chemineer HE-3 impeller showed fluctuations of much longer periods than those of the PBT.

Roussinova et al. (2000, 2001) performed LDA measurements in two tank sizes (T = 0.24 and 1.22 m), using various impeller types, impeller sizes, clearances, number of baffles (2 and 4) and working fluids in fully turbulent regime. For a 45° PBT of D = T/2 they observed a macro-instability non-dimensional frequency of $f' = 0.186$. Such frequency was coherent as the PBT was set at C = 0.25T, and such a configuration was called "resonant" geometry, whereas a broad low frequency band was observed for different clearances. The same authors performed also a LES of the vessel stirred by a PBT and confirmed the above non dimensional frequency value. In a later work Roussinova et al. (2003) identified three possible mechanisms triggering the above flow instabilities: the impingement of the jet-like impeller stream on either the vessel walls or bottom, converging radial flow at the vessel bottom from the baffles and shedding of trailing vortices from the impeller blades. For the resonant geometry, the first mechanism coincided with the impingement of the discharge stream on the vessel corner, generating pressure waves reflected back towards the impeller. The impingement jet frequency was approached with a dimensional analysis based on the Strouhal number. We well denote such flow instabilities as jet impingement instabilities (J-MIs). In a later investigation Roussinova et al. (2004) extended the analysis to different axial impellers. In such work the authors used the LSP method for the spectral analysis.

Paglianti et al. (2006) analysed literature data on MIs as well as comprehensive data obtained from measurements of wall pressure time series, and develop a simple model (based on a flow number) for predicting the MI frequency due to impinging jets (J-MI).

Also Galletti et al. (2005b) investigated flow instability for a PBT and detected a f′ = 0.187 (thus akin the Roussinova et al., 2003). Such instabilities were found to prevail in the region close to the impeller (just above it and below it in the discharged direction).

Nurtono et al. (2009) found a similar frequency f′ = 0.185 from LES modelling of a D=T/3 PBT placed at C = T/2 at Re = 40,000.

The LES results on different impellers (DT, PBTD60, PBTD45, PBTD30 and HF) from Murthy & Joshi (2008) showed the presence of J-MIs with f′ = 0.13-0.2. Moreover they observed a frequency f′ = 0.04–0.07, which lies in between the precessional and the jet instability; such frequency was attributed to the interaction of precessing vortex instability with either the mean flow or jet/circulation instabilities.

Roy et al. (2010) investigated through both experimental (PIV) and numerical (LES) techniques, the flow induced by a PBT impeller at different Reynolds numbers (Re = 44,000, 88,000 and 132,000). They found low frequency flow instabilities with frequencies of about f′ = 0.2. They could not resolve lower frequencies because of the short observation (due to computational cost of LES models) of their simulations. The authors showed changes in the three-dimensional flow pattern during different phases of the macro-instability cycle. They concluded that one mechanism driving flow instabilities was the interaction of the impeller jet stream with the tank baffles. The flow-instabilities were also observed to affect the dynamics of trailing edge vortices.

More recently Galletti & Brunazzi (2008) investigated through LDA and flow visualisation the flow features of an unbaffled vessel stirred by an eccentrically positioned Rushton turbine. The flow field evidenced two main vortices: one departing from above the impeller towards the top of the vessel and one originating from the impeller blades towards the vessel bottom. The former vortex was observed to dominate all vessel motion, leading to a strong circumferential flow around it.

The frequency analysis of LDA data indicated the presence of well defined peaks in the frequency spectra of velocity recordings. In particular three characteristic frequencies were observed in different locations across the vessel: f′ = 0.105, 0.155 and 0.94. Specifically, the f′ = 0.155 and 0.105 frequencies were related to the periodic movements of the upper and lower vortices' axis, respectively, which are also well visible from flow visualization experiments (see Fig. 6a and Fig. 6b, respectively). The f′ = 0.94 frequency was explained by considering the vortical structure – shaft interaction, which occurs in eccentric configuration and leads to vortex shedding phenomena. The authors provided an interpretation based on the Strouhal number.

In a later work (Galletti et al., 2009) the effect of blade thickness t_b was investigated, finding that for a thicker impeller (t_b/D = 0.05) the frequency of the upper vortex movement was

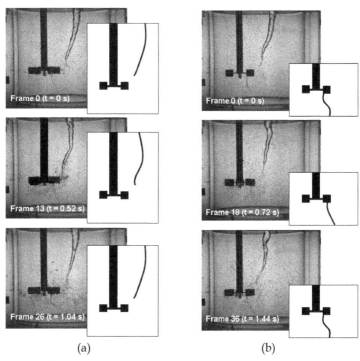

(a)   (b)

Fig. 6. Frames taken from flow visualisation experiments with sketches at N = 400 rpm (from Galletti & Brunazzi, 2008). Unbaffled vessel, RT, eccentricity E/T =0.21, C/T= 0.33, , D/T = 0.33, t_b/D = 0.01.

lower, i.e. f′ = 0.143 than for the thinner one (f′ = 0.155 for t_b/D = 0.01). The origin of the above instabilities in not fully clarified. The frequencies are one order of magnitude higher than the P-MIs frequencies. The values of f′ found are more similar to frequencies typical of J-MIs. Actually the eccentric position of the shaft and the consequently reduced distance between the impeller blade tip and the vessel boundaries, is likely to enhance the strength of the impeller discharged stream – wall interaction. In such a case, resulting flow instabilities will show a frequency which is expected to increase with increasing the velocity of the impeller discharged stream (see the flow-instability analysis in terms of pumping number by Paglianti et al., 2006, and/or peak velocity by Roussinova et al., 2003), thus with decreasing the blade thickness (Rutherford et al., 1996b).

5. Effect of flow instabilities

Flow instabilities may affect mixing operations in mechanically agitated vessels in different manners.

Since the energetic content of J-MIs may be significant, they can exert strong forces on the solid surfaces immersed in the stirred tank, i.e. the shaft, baffles, heating and cooling coils, etc. (Hasal et al., 2004). These forces may cause mechanical failure of the equipment and therefore they should be taken into account in the design of industrial-scale stirred vessels.

However except for such drawbacks, MIs may be beneficially utilized to improve mixing, provided that their phenomenology is well understood.

It has been proved than flow instabilities in stirred vessels can have a direct effect on overall parameters, which are fundamental for the design practice. The different studies on the change of circulation pattern (mentioned in section 4.1) have evidenced that such change is accompanied by a change of power number. In case of solid suspension, changes in the Njs is observed. Thus the knowledge of parameters affecting the circulation change may help optimising solid-liquid operations. Moreover, the heat flux studies of Haam et al. (1992) showed that precessional MIs may induce a variation of the heat transfer coefficient up to 68% near the surface.

Macro-instabilities may have beneficial implications for mixing process operation and efficiency as such flow motions can enhance mixing through mean-flow variations. For example, the associated low-frequency, high-amplitude oscillatory motions in regions of low turbulence in a vessel, have the capability of transporting substances fed to a mixing process over relatively long distances, as demonstrated by Larsson et al. (1996). These authors measured glucose concentration in a cultivation of Saccharomyces Cerevisiae and observed fluctuations of glucose concentration which were more pronounced as the feed was located in a stagnant area rather than in the well-mixed impeller area. Therefore flow instabilities may help destroying segregated zones inside the tank. Ducci & Yianneskis (2007) showed that the mixing time could be reduced even by 30% if the tracer is inserted at or near the MI vortex core. Houcine et al. (1999) reported with LIF a feedstream jet intermittency in a continuous stirred tank reactor due to MIs. Recently also Galletti et al. (2009) observed from decolourisation experiments in an eccentrically agitated unbaffled vessel that the flow instability oscillations help the transport of reactants far away if these are fed in correspondence of the vortices shown in Fig. 6.

Subsequently MIs have similar effects to those reported for laminar mixing in stirred tanks by Murakami et al. (1980), who observed that additional raising and lowering of a rotating impeller produced unsteady mean flow motions that either destroyed segregated regions or prevented them from forming, and could produce desired mixing times with energy savings of up to 90% in comparison to normal impeller operation. Later Nomura et al. (1997) observed that the reversal of the rotational direction of an impeller could also decrease mixing times as the additional raising or lowering of the impeller.

For a solid-liquid system (solid volume fractions up to 3.6%) agitated by a D = T/3 RT in turbulent regime (Re = 100,000 and 150,000) Derksen (2003) showed that the precessing vortex may help the resuspension of particles lying on the bottom of the tank, thus enhancing the mass transfer.

Guillard et al. (2000a) carried out LIF experiments on a stirred tank equipped with two RT observing large time scale oscillations of the concentration, induced by an interaction between the flows from the impeller and a baffle. They argued that circulation times can be altered when the flow direction changes, the turbulence levels measured with stationary probes can be significantly broadened and thus can provide an erroneous interpretation of the true levels of turbulence in a tank, and mixing in otherwise quiescent regions can be significantly enhanced due to the presence of flow variations (Guillard et al., 2000b). Knoweledge of true levels of turbulence is needed for the optimum design of micro-mixing operations (as in cases of chemical reactions). Also Nikiforaki et al. (2003) observed that P-MIs can broaden real turbulence levels up to 25% for a PBT.

Actually the problem is rather complex as Galletti et al. (2005b) as well other investigators (e.g. Ducci & Yianneskis, 2007, Roussinova et al., 2004) showed that different kinds of macro-instabilities may be present simultaneously in stirred vessels. For instance Galletti et al. (2005b) studied simultaneously with 2-point LDA the combined effect of precessional MIs and flow instabilities stemming from impeller clearance variations (CIs) in different regions of a vessel stirred with a RT. Table 1 summarizes the flow instability characteristics. The authors removed from the total energetic content of a LDA signal, the contribution of blade passage, P-MIs and CIs, evaluating the real turbulent energy. They found that the occurrence and energetic content of P-MIs and CIs depend on both measurement location and flow regime. In particular, near the vessel surface P-MIs are stronger, with energetic contents that reach 50% of the turbulent energy, meaning that they can broaden turbulence levels up to 22%. In the vicinity of the impeller the energetic content of the P-MIs is smaller, whereas CIs contribute strongly to the fluid motion with average energetic contents of about 21% of the turbulent energy for the transitional regime. Results are summarised in Table 2.

Rushton turbine		
Flow instability	CIs	P-MIs
How they manifest	change in circulation	large temporal and spatial fluctuation superimposed on the mean flow pattern
Impeller/vessel configuration	specific configuration (C/T = 0.17-0.2 with D/T = 0.33)	several configurations (different impeller types D/T, C/T)
Temporal appearance	intermittently present	continuously present
Non-dimensional frequency	f' = 0.13	f' = 0.015
Possible origin	interaction between impeller discharged stream and vessel base/walls	precessional motion of a vortex about the shaft

Table 1. Characteristics of CIs and MIs investigated with the Rushton turbine. Galletti (2005).

	Near the surface		Impeller region	
	E_{MI}/E_{TUR}	E_{CI}/E_{TUR}	E_{MI}/E_{TUR}	E_{CI}/E_{TUR}
double-loop	up to 50%	~4%	~5%	~3%
transitional state	up to 25%	up to 25%	negligible	~ 21%
single-loop	~ 12 %	~ 3%	negligible	negligible

Table 2. Relative energy of MIs and CIs with respect to the turbulent energy for the double-, single- and transitional patterns. Galletti (2005).

A similar analysis was carried out for a PBT: in this case the P-MIs and J-MIs were studied (see Table 3). The authors found the presence of both instabilities, indicating that the occurrence and magnitude, i.e. energetic content, of MIs and JIs vary substantially from one region of a vessel to another. P-MIs affect strongly the region of the vessel near the surface and around the shaft, whereas the bulk of the vessel is dominated more by J-MIs generated from the interaction of the impeller discharged stream and the vessel boundaries. J-MIs are also stronger upstream of the baffles and near the walls, which may confirm their origin. Table 4 reports the energetic contribution of the different macro-instabilities at different axial location in the vessel.

Pitched Blade Turbine		
Flow instability	J-MIs	P-MIs
How they manifest	large temporal and spatial fluctuation superimposed on the mean flow pattern	large temporal and spatial fluctuation superimposed on the mean flow pattern
Impeller/vessel configuration	specific configuration ($C/T = 0.25$ with $D/T = 0.5$)	several configurations (different impeller types D/T, C/T)
Temporal appearance	continuously present	continuously present
Non-dimensional frequency	$f' = 0.186$	$f' = 0.015$
Possible origin	interaction between impeller discharged stream and vessel base/walls	precessional motion of a vortex about the shaft

Table 3. Characteristics of JIs and MIs investigated with the pitched blade turbine. Galletti (2005).

Location of the horizontal plane	P-MIs		J-MIs	
	Max E_{MI}/E_{TUR}	Average E_{MI}/E_{TUR}	Max E_{JI}/E_{TUR}	Average E_{JI}/E_{TUR}
$z/T = 0.05$	1.9%	5.7%	2.7%	6.3%
$z/T = 0.6$	6.2%	12%	10.1%	20%
$z/T = 0.93$	14.6%	39.8%	1.7%	7%

Table 4. Average and maximum relative energy of MIs and JIs with respect to the turbulent energy, for different horizontal planes. Galletti (2005).

For the eccentric agitation in an unbaffled vessel, Galletti & Brunazzi (2008) showed that the flow instability related to the movement of the two vortices described in section 4.2.2. was very strong, as its energetic contribution was evaluated to be as high as 52% of the turbulent kinetic energy. Also the shedding vortices from flow-shaft interaction considerably affected the turbulence levels (energetic contribution of 82%), hence they should be considered in evaluating the micro-mixing scales.

6. References

Armenante, P.M. & Nagamine, E.U. (1998). Effect of low off-bottom impeller clearance on the minimum agitation speed for complete suspension of solids in stirred tanks. *Chemical Engineering Science,* Vol. 53, pp. 1757-1775.

Assirelli, M.; Bujalski, W.; Eaglesham, A. & Nienow, A.W. (2005). Intensifying micromixing in a semi-batch reactor using a Rushton turbine. *Chemical Engineering Science,* Vol. 60, pp.2333-2339.

Bakker, A.; Fasano, J.B. & Myers, K.J. (1998). Effects of flow pattern on the solid distribution in a stirred tank, Published in *"The online CFM Book"* at *http://www.bakker.org.cfm.*

Bakker, A.; La Roche, R.D.; Wang, M. & Calabrese, R. (1997). Sliding mesh simulation of laminar flow in stirred reactors. *Transactions IChemE, Chemical Engineering Research and Design,* Vol. 75, pp. 42-44.

Baldi, G.; Conti, R. & Alaria, E. (1978). Predicting the minimum suspension speeds in agitated tanks. *Chemical Engineering Science,* Vol. 33, pp. 21-25.

Berkooz, G.; Holmes, P. & Lumley, J.L. (1993). The proper orthogonal decomposition in the analysis of turbulent flows. *Annual Review of Fluid Mechanics,* Vol. 25, pp. 539–576.

Bruha, O.; Fort, I. & Smolka, P. (1995). Phenomenon of turbulent macro-instabilities in agitated systems. *Collection of Czechoslovak Chemical Communications,* Vol. 60, pp. 85-94.

Bruha, O.; Fort, I.; Smolka, P. & Jahoda, M. (1996). Experimental study of turbulent macroinstabilities in an agitated system with axial high-speed impeller and with radial baffles. *Collection of Czechoslovak Chemical Communications,* Vol. 61, pp. 856-867.

Conti, R.; Sicardi, S. & Specchia, V. (1981). Effect of the stirrer clearance on suspension in agitated vessels. *Chemical Engineering Journal,* Vol. 22, pp. 247-249.

Cooley, J.W. & Tukey, J.W. (1965). An algorithm for the machine calculation of complex Fourier series. *Mathematical Computation,* Vol. 19, pp. 297-301.

Daubechies, I. (1990). The Wavelet transform time-frequency localization and signal analysis. *IEEE Transactions on Information Theory,* Vol. 36, pp. 961-1005.

Derksen, J.J. (2003). Numerical simulation of solids suspension in a stirred tank, *AIChE Journal.* Vol. 49, pp. 2700-2714.

Distelhoff, M.F.W.; Laker, J.; Marquis, A.J. & Nouri, J. (1995). The application of a strain-gauge technique to the measurement of the power characteristics of 5 impellers. *Experiments in Fluids,* Vol. 20, pp. 56-58.

Doulgerakis, Z.; Yianneskis, M.; Ducci, A. (2011). On the interaction of trailing and macro-instability vortices in a stirred vessel-enhanced energy levels and improved mixing potential. *Chemical Engineering Research and Design,* Vol. 87, pp. 412-420.

Ducci, A. & Yianneskis, M. (2007). Vortex tracking and mixing enhancement in stirred processes. *AIChE Journal,* Vol. 53, pp. 305-315.

Ducci, A.; Doulgerakis, Z. & Yianneskis, M. (2008). Decomposition of flow structures in stirred reactors and implications for mixing enhancement. *Industrial and Engineering Chemistry Research,* Vol. 47, pp. 3664-3676.

Galletti, C. & Brunazzi, E. (2008). On the main flow features and instabilities in an unbaffled vessel agitated with an eccentrically located impeller. *Chemical Engineering Science,* Vol 63, pp. 4494-4505

Galletti, C. (2005). *Experimental analysis and modeling of stirred vessels*. Ph.D. thesis, University of Pisa, Pisa, Italy.

Galletti, C.; Brunazzi, E.; Pintus, S.; Paglianti, A. & Yianneskis, M. (2004a). A study of Reynolds stresses, triple products and turbulence states in a radially stirred tank with 3-D laser anemometry. *Transactions IChemE, Chemical Engineering Research and Design,* Vol. 82, pp. 1214-1228.

Galletti, C.; Brunazzi, E.; Yianneskis, M. & Paglianti, A. (2003). Spectral and Wavelet analysis of the flow pattern transition with impeller clearance variations in a stirred vessel. *Chemical Engineering Science,* Vol. 58, pp. 3859-3875.

Galletti, C.; Lee, K.C.; Paglianti, A. & Yianneskis, M. (2005a). Flow instabilities associated with impeller clearance changes in stirred vessels. *Chemical Engineering Communications,* Vol. 192, pp. 516-531.

Galletti, C.; Lee, K.C.; Paglianti, A. & Yianneskis, M. (2004b). Reynolds number and impeller diameter effects on instabilities in stirred vessels, *AIChE Journal,* Vol. 50, 2050-2063.

Galletti, C.; Paglianti, A. & Yianneskis, M. (2005b). Observations on the significance of instabilities, turbulence and intermittent motions on fluid mixing processes in stirred reactors. *Chemical Engineering Science,* Vol. 60, pp. 2317-2331.

Galletti, C.; Pintus, S. & Brunazzi, E. (2009). Effect of shaft eccentricity and impeller blade thickness on the vortices features in an unbaffled vessel. *Chemical Engineering Research and Design,* Vol. 87, pp. 391-400.

Guillard, F.; Trägårdh, C. & Fuchs, L. (2000a). A study of turbulent mixing in a turbine-agitated tank using a fluorescence technique. *Experiments in Fluids,* Vol. 28, pp. 225-235.

Guillard, F.; Trägårdh, C. & Fuchs, L. (2000b). A study on the instability of coherent mixing structures in a continuously stirred tank. *Chemical Engineering Science,* Vol. 55, pp. 5657-5670.

Haam, S.; Brodkey, R.S. & Fasano, J.B. (1992). Local heat-transfer in a mixing vessel using heat-flux sensors. *Industrial and Engineering Chemistry Research,* Vol. 31, 1384-1391.

Hartmann, H.; Derksen, J.J. & van den Akker H.E.A. (2006). Numerical simulation of a dissolution process in a stirred tank reactor. *Chemical Engineering Science.* Vol 61 , pp. 3025 – 3032

Hartmann, H.; Derksen, J.J. & van den Akker, H.E.A. (2004). Macroinstability uncovered in a Rushton turbine stirred tank by means of LES. *AIChE Journal,* Vol. 50, pp. 2383-2393.

Hasal, P.; Fort, I. & Kratena, J. (2004). Force effects of the macro-instability of flow pattern on radial baffles in a stirred vessel with pitched-blade and Rushton turbine impeller. *Transactions IChemE, Chemical Engineering Research and Design,* Vol. 82, pp. 1268-1281.

Hasal, P.; Montes, J.L.; Boisson, H.C. & Fort, I. (2000). Macro-instabilities of velocity field in stirred vessel: detection and analysis. *Chemical Engineering Science,* Vol. 55, pp. 391-401.

Hockey, R. M. & Nouri, M. (1996). Turbulent flow in a baffled vessel stirred by a 60° pitched blade impeller. *Chemical Engineering Science,* Vol. 51, pp. 4405-4421.

Hockey, R.M. (1990). *Turbulent Newtonian and non-Newtonian flows in a stirred reactor*, Ph.D. thesis, Imperial College, London.

Houcine, I.; Plasari, E.; David, R. & Villermaux, J. (1999). Feedstream jet intermittency phenomenon in a continuous stirred tank reactor. *Chemical Engineering Journal,* Vol. 72, pp. 19-29.

Ibrahim, S. & Nienow, A.W. (1995). Power curves and flow patterns for a range of impellers in Newtonian fluid: $40<Re<5 \times 10^5$. *Transactions IChemE, Chemical Engineering Research and Design,* Vol. 73, pp. 485-491.

Ibrahim, S. & Nienow, A.W. (1996). Particle suspension in the turbulent regime: the effect of impeller type and impeller/vessel configuration. *Transactions IChemE, Chemical Engineering Research and Design,* Vol. 74, pp. 679-688.

Jaworski, Z.; Nienow, A.W.; Koutsakos, E.; Dyster, K. & Bujalski, W. (1991). An LDA study of turbulent flow in a baffled vessel agitated by a pitched blade turbine. *Transactions IChemE, Chemical Engineering Research and Design,* Vol. 69, pp. 313-320.

Kilander, J; Svensson, F.J.E. & Rasmuson, A. (2006). Flow instabilities, energy levels, and structure in stirred tanks. *AIChE Journal,* Vol. 52, pp. 3049-4051.

Kresta, S.M. & Wood, P.E. (1993). The mean flow field produced by a 45° pitched-blade turbine: changes in the circulation pattern due to off bottom clearance. *Canadian Journal of Chemical Engineering,* Vol. 71, pp. 42-53.

Larsson G.; To¨rnkvist M.; Ståhl Wernersson E.; Tra¨gårdh ,C.; Noorman, H. & Enfors, S.O. (1996). Substrate gradients in bioreactors: origin and consequences. *Bioprocess Engineering,* Vol. 14, pp. 281-289.

Lavezzo, V.; Verzicco, R. & Soldati, A. (2009). Ekman pumping and intermittent particle resuspension in a stirred tank reactor. *Chemical Engineering Research and Design.* Vol. 87, pp. 557–564.

Lomb, N.R. (1976). Least-squares frequency analysis of unequally spaced data, *Astrophysics and Space Science.* Vol. 39, pp. 447-462.

Mao, D.; Feng, L.; Wang, K. & Li, Y. (1997). The mean flow generated by a pitched-blade turbine: changes in the circulation pattern due to impeller geometry, *Canadian Journal of Chemical Engineering,* Vol. 75, pp. 307-316.

Micheletti, M. & Yianneskis, M. (2004). Precessional flow macro-instabilities in stirred vessels: study of variations in two locations through conditional phase-averaging and cross-correlation approaches, *Proceedings 11th International Symposium on Applications of Laser Techniques to Fluid Mechanics,* Lisbon, Portugal.

Montante, G.; Lee, K.C.; Brucato, A. & Yianneskis, M. (1999). Double- to single- loop flow pattern transition in stirred vessels. *Canadian Journal of Chemical Engineering,* Vol. 77, pp. 649-659.

Montes, J.L.; Boisson H.C.; Fort, I. & Jahoda, M. (1997). Velocity field macro-instabilities in an axially agitated mixing vessel. *Chemical Engineering Journal,* Vol. 67, pp. 139-145.

Murakami, Y.; Hirose, T.; Yamato, T.; Fujiwara, H. & Ohshima, M. (1980). Improvement in mixing of high viscosity liquid by additional up-and-down motion of a rotating impeller. *Journal of Chemical Engineering of Japan,* Vol. 13, pp. 318-323.

Murthy, B.N. & Joshi, J.B. (2008). Assessment of standard k–ε, RSM and LES turbulence models in a baffled stirred vessel agitated by various impeller designs. *Chemical Engineering Science,* Vol. 63, pp. 5468-5495

Myers, K.J.; Bakker, A. & Corpstein, R.R. (1996). The effect of flow reversal on solids suspension in agitated vessels. *Canadian Journal of Chemical Engineering* 74, pp. 1028-1033.

Myers, K.J.; Ward, R.W. & Bakker, A. (1997). A digital particle image velocimetry investigation of flow field instabilities of axial-flow impellers. *Journal of Fluid Engineering*, Vol. 119, pp. 623-632.

Nienow, A. W. (1968). Suspension of solid particles in turbine agitated baffled vessels. *Chemical Engineering Science*, Vol. 23, pp. 1453-1459.

Nikiforaki, L.; Montante, G.; Lee, K.C.; Yianneskis, M. (2003). On the origin, frequency and magnitude of macro-instabilities of the flows in stirred vessels. *Chemical Engineering Science*, Vol. 58, pp. 2937-2949.

Nikiforaki, L.; Yu, J.; Baldi, S.; Genenger, B.; Lee, K.C.; Durst, F. & Yianneskis, M. (2004). On the variation of precessional flow instabilities with operational parameters in stirred vessels. *Chemical Engineering Journal*, Vol. 102, pp. 217-231.

Nomura, T.; Uchida, T. & Takahashi, K. (1997). Enhancement of mixing by unsteady agitation of an impeller in an agitated vessel. *Journal of Chemical Engineering of Japan*, Vol. 30, pp. 875-879.

Nouri, J.M. & Whitelaw, J.H. (1990). Flow characteristics of stirred reactors with Newtonian and non-Newtonian fluids. *AIChE Journal*, Vol. 36, pp. 627-629.

Nurtono T.; Setyawan; H. Altway A. & Winardi, S. (2009) Macro-instability characteristic in agitated tank based on flow visualization experiment and large eddy simulation. *Chemical Engineering Research and Design*, Vol. 87, pp. 923-942.

Paglianti, A.; Montante, G. & Magelli, F. (2006). Novel experiments and a mechanistic model for macroinstabilities in stirred tanks. *AIChE Journal*, Vol. 52, pp. 426-437.

Roussinova, V.T.; Grgic, B. and Kresta, S.M. (2000). Study of macro-instabilities in stirred tanks using a velocity decomposition technique. *Transactions IChemE, Chemical Engineering Research and Design*, Vol. 78, 1040-1052.

Roussinova, V.T.; Kresta, S.M. & Weetman, R. (2003). Low frequency macroinstabilities in a stirred tank: scale-up and prediction based on large eddy simulations. *Chemical Engineering Science*, Vol. 58, pp. 2297-2311.

Roussinova, V.T.; Kresta, S.M. & Weetman, R. (2004). Resonant geometries for circulation pattern macroinstabilities in a stirred tank. *AIChE Journal*, Vol. 50, pp. 2986-3005.

Roussinova, V.T.; Weetman, R. & Kresta, S.M. (2001). Large eddy simulation of macro-instabilities in a stirred tank with experimental validation at two scales. *North American Mixing Forum 2001*.

Roy, S.; Acharya, S. & Cloeter, M. (2010). Flow structure and the effect of macro-instabilities in a pitched-blade stirred tank. *Chemical Engineering Science*, Vol. 65, pp. 3009–3024.

Rutherford, K.; Lee, K.C.; Mahmoudi, S.M.S. & Yianneskis, M. (1996a). Hydrodynamic characteristics of dual Rushton impeller stirred vessels. *AIChE Journal*, Vol. 42, pp. 332-346.

Rutherford, K.; Mahmoudi, S.M.S.; Lee, K.C. & Yianneskis, M. (1996b). The Influence of Rushton impeller blade and disk thickness on the mixing characteristics of stirred vessels. *Transactions IChemE, Chemical Engineering Research and Design*, Vol. 74, pp. 369-378.

Scargle, J. (1982). Studies in astronomical time series analysis. II Statistical aspect of spectral analysis of unevenly spaced data. *Astrophysical Journal*, Vol. 263, pp. 835-853.

Schäfer, M.; Yianneskis, M.; Wächter, P. & Durst, F. (1998). Trailing vortices around a 45° pitched-blade impeller. *AIChE Journal*, Vol. 44, pp. 1233-1246.

Sharma, R.N. & Shaikh, A.A. (2003). Solids suspension in stirred tanks with pitched blade turbines. *Chemical Engineering Science,* Vol. 58, pp. 2123-2140.

Tatterson, G.B. (1991). *Fluid mixing and gas dispersion in agitated tanks,* Mcgraw-Hill Inc., New York.

Tiljander, P. & Theliander, H. (1993). Power-consumption and solid suspension in completely filled vessels. *Chemical Engineering Communications,* Vol. 124, pp. 1-14.

Torré, J.P.; Fletcher, D.F; Lasuye T. & Xuereb, C. (2007) Single and multiphase CFD approaches for modelling partially baffled stirred vessels: Comparison of experimental data with numerical predictions. *Chemical Engineering Science,* Vol. 62, pp. 6246-6262

Torrence, C. & Compo, P. (1998). A practical guide to Wavelet analysis. *Bulletin of the American Meteorological Society,* Vol. 79, pp. 61-78.

Yianneskis, M.; Popiolek, Z. & Whitelaw, J.H. (1987). An experimental study of the steady and unsteady flow characteristics of stirred reactors. *Journal of Fluid Mechanics,* Vol. 175, pp. 537-555.

Zwietering, T.N. (1958). Suspending of solid particles in liquid by agitators. *Chemical Engineering Science,* Vol. 8, pp. 244-253.

9

Microrheology of Complex Fluids

Laura J. Bonales, Armando Maestro, Ramón G. Rubio and Francisco Ortega
Departamento de Química Física I, Facultad de Química,
Universidad Complutense, Madrid
Spain

1. Introduction

Many of the diverse material properties of soft materials (polymer solutions, gels, filamentous proteins in cells, etc.) stem from their complex structures and dynamics with multiple characteristic length and time scales. A wide variety of technologies, from paints to foods, from oil recovery to processing of plastics, all heavily rely on the understanding of how complex fluids flow (Larson, 1999).

Rheological measurements on complex materials reveal viscoelastic responses which depend on the time scale at which the sample is probed. In order to characterize the rheological response one usually measures the shear or the Young modulus as a function of frequency by applying a small oscillatory strain of frequency ω. Typically, commercial rheometers probe frequencies up to tens of Hz, the upper range being limited by the onset of inertial effects, when the oscillatory strain wave decays appreciably before propagating throughout the entire sample. If the strain amplitude is small, the structure is not significantly deformed and the material remains in equilibrium; in this case the affine deformation of the material controls the measured stress, and the time-dependent stress is linearly proportional to the strain (Riande et al., 2000).

Even though standard rheological measurements have been very useful in characterizing soft materials and complex fluids (e.g. colloidal suspensions, polymer solutions and gels, emulsions, and surfactant solutions), they are not always well suited for all systems because milliliter samples are needed thus precluding the study of rare or precious materials, including many biological samples that are difficult to obtain in large quantities. Moreover, conventional rheometers provide an average measurement of the bulk response, and do not allow for local measurements in inhomogeneous systems. To address these issues, a new methodology, microrheology, has emerged that allows to probe the material response on micrometer length scales with microliter sample volumes. Microrheology does not correspond to a specific experimental technique, but rather a number of approaches that attempt to overcome some limitations of traditional bulk rheology (Squires & Mason, 2010; Wilson & Poon, 2011). Advantages over macrorheology include a significantly higher range of frequencies available without time-temperature superposition (Riande et al., 2000), the capability of measuring material inhomogeneities that are inaccessible to macrorheological methods, and rapid thermal and chemical homogeneization that allow the transient rheology of evolving systems to be studied (Ou-Yang & Wei, 2010). Microrheology methods typically use embedded micron-sized probes to locally deform the sample, thus allowing one to use this type of rheology on very small volumes, of the order of a microliter. Macro-

and microrheology probe different aspects of the material: the former makes measurements over extremely long (macroscopic) length scales using a viscometric flow field, whereas the latter effectively measures material properties on the scale of the probe itself (Squires & Mason, 2010; Breedveld & Pine, 2003). As the probe increases in size, one might expect that micro- and macrorheology would converge, however, as it has been suggested, it is possible that macro- and microrheology techniques do not probe exactly the same physical properties because - even in the continuum (large probe) limit - one experiment uses a viscometric flow whereas the other does not (Kahir & Brady, 2005; Lee et al., 2010; Schmidt et al., 2000; Oppong & de Bruyn, 2010).

One can distinguish two main families of microrheological experiments: One type of experiments focuses on the object itself; for example, the study of motor proteins aims at understanding the mechanical motions of the protein associated with enzymatic activities on the molecular level (Ou-Yang & Wei, 2010). The other type of experiment aims at understanding the local environment of the probe by observing changes in its random movements (Crocker & Grier, 1996; MacKintosh & Schmidt, 1999). Fundamentally different from relaxation kinetics, microrheology measures spontaneous thermal fluctuations without introducing major external perturbations into the systems being investigated. Other well-established methods in this family are dynamic light scattering (Dasgupta et al., 2002; Alexander & Dalgleish, 2007; Tassieri et al. 2010), and fluorescence correlation spectroscopy (Borsali & Pecora, 2008; Wöll et al., 2009). With recent advancement in spatial and temporal resolution to subnanometer and submillisecond, particle tracking experiments are now applicable to study of macromolecules (Pan et al., 2009) and intracellular components such as cytoskeletal networks (Cicuta & Donald, 2007). Detailed descriptions of the methods and applications of microrheology to the study of bulk systems have been given in review articles published in recent years (Crocker & Grier 1996; MacKintosh & Schmidt, 1999; Mukhopadhyay & Granick, 2001; Waigh, 2005; Gardel et al., 2005; Cicuta & Donald, 2007).

Interfaces play a dominant role in the behavior of many complex fluids. Interfacial rheology has been found to be a key factor in the stability of foams and emulsions, compatibilization of polymer blends, flotation technology, fusion of vesicles, etc. (Langevin, 2000). Also, proteins, lipids, phase-separated domains, and other membrane-bound objects diffuse in the plane of an interface (Cicuta et al., 2007). Particle-laden interfaces have attracted much attention in recent years because of the tendency of colloidal particles to become (almost irreversibly) trapped at interfaces and their behavior once there has lead to their use in a wide variety of systems including drug delivery, stabilization of foams and emulsions, froth, flotation, or ice cream production. There still is a need to understand the colloidal interactions to have control over the structure and therefore the properties of the particle assemblies formed, specially because it has been pointed out that the interactions of the particles at interfaces are far more complex than in the bulk (Binks & Horozov, 2006; Bonales et al., 2011). In recent years books and reviews of particles at liquid interfaces have been published (Kralchewski & Nagayama, 2001). The dynamic properties of particle-laden interfaces are strongly influenced by direct interparticle forces (capillary, steric, electrostatic, van der Waals, etc.) and complicated hydrodynamic interactions mediated by the surrounding fluid. At macroscopic scales, the rheological properties of particle-laden fluid interfaces can be viewed as those of a liquid-liquid interface with some effective surface viscoelastic properties described by effective shear and compressional complex viscoelastic moduli.

A significant fact is that for the simplest fluid-fluid interface, different dynamic modes have to be taken into account: the capillary (out of plane) mode, and the in-plane mode, which contains dilational (or extensional) and shear contributions. For more complex interfaces, such as thicker ones, other dynamic modes (bending, splaying) have to be considered (Miller & Liggieri, 2009). Moreover, the coupling of the abovementioned modes with adsorption/desorption kinetics may be very relevant for interfaces that contain soluble or partially soluble surfactants, polymers or proteins (Miller & Liggieri, 2009; Muñoz et al., 2000; Díez-Pascual et al. 2007). In the case of surface shear rheology, most of the information available has been obtained using macroscopic interfacial rheometers which in many cases work at low Boussinesq numbers (Barentin et al., 2000; Gavranovic et. al., 2005; Miller & Liggieri, 2009; Maestro et al., 2011.a). Microrheology has been foreseen as a powerful method to study the dynamics of interfaces. In spite that the measurement of diffusion coefficients of particles attached to the interface is relatively straightforward with modern microrheological techniques, many authors have relied on hydrodynamic models of the viscoelastic surroundings traced by the particles in order to obtain variables such as interfacial elasticity or shear viscosity. The more complex the structure of the interface the stronger are the assumptions of the model, and therefore it is more difficult to check their validity. In the present work we will briefly review modern microrheology experimental techniques, and some of the recent results obtained for bulk and interfacial systems. Finally, we will summarize the theoretical models available for calculating the shear microviscosity of fluid monolayers from particle tracking experiments, and discuss the results for some systems.

2. Experimental techniques

For studying the viscoelasticity of the probe environment there are two broad types of experimental methods: active methods, which involve probe manipulation, and passive methods, that relay on thermal fluctuations to induce motion of the probes. Because thermal driving force is small, no sample deformation occurs that exceeds equilibrium thermal fluctuations. This virtually guarantees that only the linear viscoelastic response of the embedding medium is probed (Waigh, 2005). On the contrary, active methods allow the nonlinear response to be inferred from the relationship between driving force and probe velocity, in such cases the microstructure itself can be deformed significantly so that the material response differs from the linear case (Squires, 2008). As a consequence, passive techniques are typically more useful for measuring low values of predominantly viscous moduli, whereas active techniques can extend the measurable range to samples with significant elasticity modulus. Figure 1 shows the typical ranges of frequencies and shear moduli that can be studied with the different microrheological techniques.

2.1 Active techniques
2.1.1 Magnetic tweezers
This is the oldest implementation of an active microrheology technique, and it has been recently reviewed by Conroy (Conroy, 2008). A modern design has been described by Keller et al. (2001). The method combines the use of strong magnets to manipulate embedded super-paramagnetic or ferromagnetic particles, with video microscopy to measure the displacement of the particles upon application of constant or time-dependent forces. Strong magnetic fields are required to induce a magnetic dipole in the beads and magnetic field gradients are applied to produce a force. The force exerted is typically in the range of 10 pN to 10 nN depending on the experimental details (Keller et al. 2001). The spatial resolution is typically in the range of 10-20 nm, and the frequency range is 0.01 – 1000 Hz. Three modes

of operation are possible: a viscosimetry measurement after applying a constant force, a creep response experiment after applying a pulse excitation, and the measurement of the frequency dependent viscoelastic moduli in response to an oscillatory stress (Riande et al., 2000). This technique has been extensively applied to characterize the bulk viscoelasticity of systems of biological relevance (Wilson & Poon, 2011; Gardel et al., 2005). Moreover, real-time measurements of the local dynamics have also been reported for systems which change in response to external stimuli (Bausch et al., 2001), and rotational diffusion of the beads has also been used to characterize the viscosity of the surrounding fluid and to apply mechanical stresses directly to the cell surfaces receptors using ligand coated magnetic colloidal particles deposited onto the cell membrane (Fabry et al., 2001). Finally, this technique is well suited for the study of anisotropic systems by mapping the strain-field, and for studying interfaces (Lee et al., 2009). In recent years (Reynaert et al., 2008) have described a magnetically driven macrorheometer for studying interfacial shear viscosities in which one of the dimensions of the probe (a magnetic needle) is in the μm range. This has allowed the authors to work at rather high values of the Boussinesq number, which is one of the typical characteristics of the microrheology techniques.

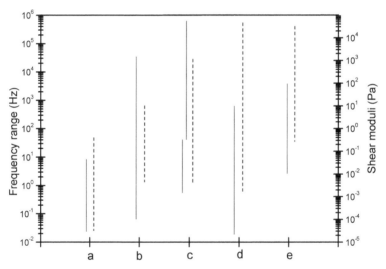

Fig. 1. Frequency and elasticity modulus range available to the different microrheological techniques. Continuous vertical represent the frequency range, and dashed arrows the range of shear moduli (G', G'') that are accessible to each technique. a)Video particle tracking. b) Optical Tweezers. c) Diffusing wave spectroscopy: upper line for transmission geometry, lower line for back geometry. d) Magnetic microrheology. e) Atomic Force Microscopy (AFM). Adapted from Waigh (2005).

2.1.2 Optical tweezers

This technique uses a highly focused laser beam to trap a colloidal particle, as a consequence of the momentum transfer associated with bending light. The most basic design of an optical tweezer is shown in Figure 2.a: A laser beam (usually in the IR range) is focused by a high-quality microscope (high numerical aperture objective) to a spot in a plane in the fluid.

Figure 2.b shows a detailed scheme of how an optical trap is created. Light carries a momentum, in the direction of propagation, that is proportional to its energy. Any change in the direction of light, by reflection or refraction, will result in a change of the momentum of the light. If an object bends the light, conservation momentum requires that the object must undergo an equal and opposite momentum change, which gives rise to a force acting on the subject. In a typical instrument the laser has a Gaussian intensity profile, thus the intensity at the center is higher than at the edges. When the light interacts with a bead, the sum of the forces acting on the particle can be split into two components: F_{sc}, the scattering force, pointing in the direction of the incident beam, and F_g, the gradient force, arising from the gradient of the Gaussian intensity profile and pointing in the plane perpendicular to the incident beam towards the center of the beam. F_g is a restoring force that pulls the bead into the center of the beam. If the contribution to F_{sc} of the refracted rays is larger than that of the reflected rays then a restoring force is also created along the beam direction and a stable trap exists. A detailed description of the theoretical basis and of modern experimental setups has been given in Refs. (Ou-Yang & Wei, 2010; Borsali & Pecora, 2008; Resnick, 2003) that also include a review of applications of optical and magnetic tweezers to problems of biophysical interest: ligand-receptor interactions, mechanical response of single chains of biopolymers, force spectroscopy of enzymes and membranes, molecular motors, and cell manipulation. A recent application of optical tweezers to study the non-linear mechanical response of red-blood cells is given by Yoon et al. (2008). Finally, optical tweezers are also suitable for the study of interfacial rheology (Steffen et al., 2001).

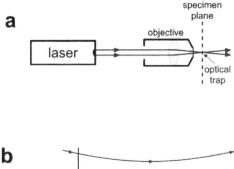

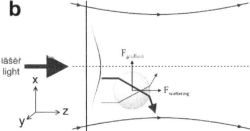

Fig. 2. a) Basic design of an optical tweezers instrument. b) Details of the physical principles leading to the optical trap.

2.2 Passive techniques

These techniques use the Brownian dynamics of embedded colloids to measure the rheology of the materials. Since passive methods use only the thermal energy of the beads, materials must be sufficiently soft for the motion of the particles to be measure precisely. The resolution typically ranges from 0.1 to 10 nm and elastic modulus from 10 to 500 Pa can be measured with micron sized particles. Thermal fluctuations of particles in transparent bulk systems have traditionally been studied using light scattering techniques that allow one to measure the intensity correlation function from which the field correlation function $g_1(t)$ can be calculated, t being the time. For monodisperse particles $g_1(t)$ is directly related to the mean squared displacement of the particles, MSD, through

$$g_1(t) = \exp{[-q^2 <\Delta r^2(t)>/6]} \qquad (1)$$

q being the scattering wave vector (Borsali & Pecora, 2008). Once $<\Delta r^2(t)>$ is obtained, it is possible to calculate the real and imaginary components of the shear moduli, G' and G" (Oppong & de Bruyn, 2010).

2.2.1 Diffusion wave spectroscopy
Diffusion wave spectroscopy, DWS, allows measurements of multiple scattering media, and therefore non-transparent samples can be studied. The output of the technique allows to calculate $<\Delta r^2(t)>$, and because of the multiple scattering all q-dependent information is lost as photons average over all possible angles, thus resulting only in two possible scattering geometries: transmission and backscattering. The frequency range of both geometries is complementary (see Figure 1) spanning from 0.1 Hz to 1MHz. For bulk polymer solutions and gels excellent agreement of the G' and G" values obtained by DWS and those obtained with conventional rheology has been found (Dasgupta et al., 2002; Dasgupta & Weitz, 2005). Even though these light scattering techniques are quite powerful tools for bulk microrheology, they have been scarcely used to probe the rheology of interfaces; in fact, as far as we know, only in old papers of Rice's group a set-up was described to measure dynamic light scattering of polymer monolayers using evanescent waves (Lin et al., 1993; Marcus et al., 1996).

2.2.2 Fluorescence correlation spectroscopy (FCS)
It is usually combined with optical microscopy, in particular confocal or two-photon microscopy. In these techniques light is focused on a sample and the fluorescence intensity fluctuations (due to diffusion, physical or chemical reactions, aggregations, etc.) can be measured in the form of a temporal correlation function. Similarly to what has been discussed in the light scattering technique, it is possible to obtain the MSD from the correlation function. In most experiments, Brownian motion drives the fluctuation of fluorescent-labeled molecules (or particles) within a well-defined element of the measurement cell. The samples have to be quite dilute, so that only few probes are within the focal spot (usually 1 – 100 molecules in one fL). Because of the tiny size of the confocal volume (approx. 0.2 fL), the measurements can be carried out in living cells or on cell membranes. In case that the interactions between two molecules wish to be studied, two options are available depending on their relative size. If their size is quite different, only one of them has to be labeled with a fluorescent dye (autocorrelation). If the diffusion coefficients of both molecules are similar, both have to be labeled with different dies (cross-correlation). A detailed description of FCS techniques and of the data analysis has recently given by Riegler & Elson (2001). Recent problems to which FCS has been applied include: dynamics of rafts in membranes and vesicles, dynamics of supramolecular

complexes, proteins, polymers, blends and micelles, electrically induced microflows, diffusion of polyelectrolytes onto polymer surfaces, normal and confined diffusion of molecules and polymers, quantum dots blinking, dynamics of polymer networks, enzyme kinetics and structural heterogeneities in ionic liquids (Winkler, 2007; Heuf et al., 2007; Ries & Schwille, 2008; Cherdhirankorn et al., 2009; Wöll et al., 2009; Guo et al., 2011). The use of microscopes makes FCS suitable for the study of the dynamics of particles at interfaces. Moreover, contrary to particle tracking techniques, it is not necessary to "see" the particles, thus interfaces with nanometer sized particles can be studied (Riegler & Elson, 2001).

2.2.3 Particle tracking techniques
The main idea in particle tracking is to introduce onto the interface a few spherical particles of micrometer size and follow their trajectories (Brownian motion) using videomicroscopy. The trajectories of the particles, either in bulk or on surfaces, allow one to calculate the mean square displacement, which is related to the diffusion coefficient, D, and the dimensions, d, in which the translational motion takes place by

$$\left\langle \Delta r^2(t) \right\rangle = \left\langle \left[\vec{r}(t_0 - t) - \vec{r}(t_0) \right]^2 \right\rangle = 2dDt^\alpha \tag{2}$$

where the brackets indicate the average over all the particles tracked, and t_0 the initial time. In case of diffusion in a purely viscous material or interface, α is equal to 1, and the usual linear relation is obtained between MSD and t. When the material or interface is viscoelastic, α becomes lower than 1 and this behavior is called sub-diffusive. It is worth noticing that sub-diffusivity can be found not only as a consequence of the elasticity of the material, but also due to particle interactions as concentration increases, an effect that is particularly important at interfaces. Anomalous diffusion is also found in many systems of biological interest where the Brownian motion of the particles is hindered by obstacles (Feder et al., 1996), or even constrained to defined regions (corralled motion) (Saxton & Jacobson, 1997). The diffusion coefficient is related to the friction coefficient, f, by the Einstein relation

$$D = \frac{k_B T}{f} \tag{3}$$

In 3D Stokes law, $f = 6\pi\eta a$, applies and for pure viscous fluids the shear viscosity, η, can be directly obtained from the diffusion coefficient of the probe particle of radius a at infinite dilution. The situation is much more complex in the case of fluid interfaces, and it will be discussed in more detail in the next section.

Figure 3 shows a sketch of a typical setup for particle tracking experiments. A CCD camera (typically 30 fps) is connected to a microscope that permits to image either the interface prepared onto a Langmuir trough, or a plane into a bulk fluid. The series of images are transferred to a computer to be analyzed to extract the trajectories of a set of particles. Figure 4.a shows typical results of MSD obtained for a 3D gel, combining DWS and particle tracking techniques which shows a very good agreement between both techniques, and illustrates the broad frequency range that can be explored. Figure 4.b shows a typical set of results for the MSD of a system of latex particles (1 µm of diameter) spread at the water/n-octane interface. The analysis of MSD within the linear range in terms of Eq. (2) allows to obtain D.

One of the experimental problems frequently found in particle tracking experiments is that the linear behavior of the MSD vs. t is relatively short. This may be due to poor statistics in calculating the average in Eq.(2), or to the existence of interactions between particles. As it will be discussed below, this may be a problem in calculating the shear modulus from the MSD. An additional experimental problem may be found when the interaction of the particles with the fluid surrounding them is very strong, which may lead to changes in its viscoelastic modulus, or when the samples are heterogeneous at the scale of particle size, a situation rather frequent in biological systems, e.g. cells (Konopka & Weisshaar, 2004), or gels (Alexander & Dalgleish, 2007), or solutions of rod-like polymers (Hasnain & Donald, 2006). In this case the so-called "two-point" correlation method is recommended (Chen et al., 2003). In this method the fluctuations of pairs of particles at a distance R_{ij} are measured for all the possible values of R_{ij} within the system. Vector displacements of individual particles are calculated as a function of lag time, t, and initial time, t_0.

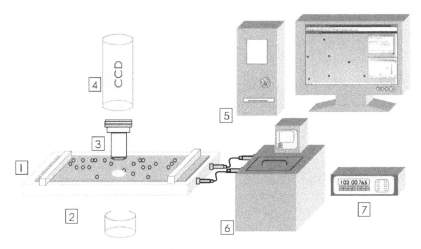

Fig. 3. Typical particle tracking setup for 2D microrheology experiments: 1: Langmuir trough; 2: illumination; 3: microscope objective; 4: CCD camera; 5: computer; 6: thermostat; 7: electronics for measuring the temperature and the surface pressure.

Then the ensemble averaged tensor product of the vector displacements is calculated (Chen et al., 2003):

$$D_{\alpha\beta}(r,\tau) = \left\langle \Delta r_\alpha^i(r,t)\Delta r_\beta^j(r,t)\delta\left[r - R_{ij}(t_0)\right]\right\rangle_{i\neq j,t} \tag{4}$$

where a and b are coordinate axes. The average corresponding to i = j represents the one-particle mean-squared displacement.

Two-point microrheology probes dynamics at different length scales larger than the particle radius, although it can be extrapolated to the particle's size thus giving the MSD (Liu et al., 2006). In fact it has been found that for R_{ij} close to the particle radius, the two-point MSD matches the tendency of the MSD obtained by tracking single particles. However, both sets of results are different for R_{ij}'s much larger than the particle diameter. This is a consequence of the fact that single particle tracking reflects both bulk and local rheologies, and therefore the heterogeneities of the sample. Figure 5 shows a comparison of the MSD obtained by single particle and two-point tracking for a solution of entangled F-actin solutions at different length scales from 1 to 100 μm (Liu et al., 2006). Both methods agree when the particle size is of the same order than the scale of the inhomogeneities of the system when the particle probes the average structure. Otherwise, the two methods lead to different results. In general, quite good agreement is found between two-point tracking experiments and macroscopic rheology experiments.

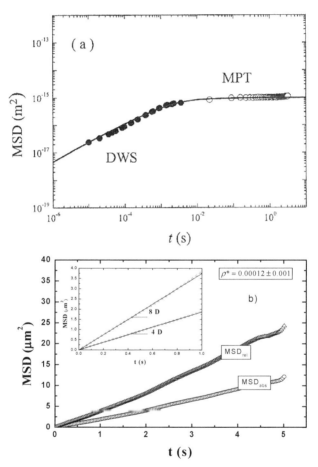

Fig. 4. a) Typical results of mean square displacement for a 3D gel made out of a polysaccharide in water [44]. Filled points are from DWS experiments, and open symbols are from particle tracking. The continuous line is an eye guide. b) Mean square displacement (MSD_{abs}), circles, and relative square displacement (MSD_{rel}), triangles, for latex particles at the water/n-octane interface. Experimental details: set of 300 latex particles of 1 μm of diameter, surface charge density: -5.8 mC cm^{-2}, and reduced surface density, ρ^*=1.2·10^{-3} (ρ^*=ρa^2), 25 °C. Figure 4.a is reproduced from Vincent et al. (2007). Inset corresponds to a smaller time interval.

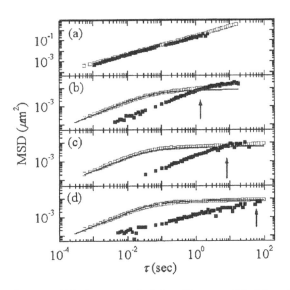

Fig. 5. Comparison of one-particle (open symbols) and two-particle (closed symbols) MSD for a solution of F-actin using particles of radius 0.42 μm. Different average actin filaments are used: a) 0.5 μm, b) 2 μm, c) 5 μm, d) 17 μm. Notice that when the scale of the inhomogeneities of the solution is similar to the particle size both methods lead to the same results. The figure is reproduced from Liu et al. (2006).

For the case in which the particles are embedded in a viscoelastic fluid, particle tracking experiments allow one to obtain the viscoelastic moduli of the fluids. Manson & Weitz (1995) first in an ad-hoc way, and later Levine & Lubensky (2000) in a more rigorous way, proposed a generalization of the Stokes-Einstein (GSE) equation:

$$\left\langle \Delta \tilde{r}^2(s) \right\rangle = \frac{2k_BT}{3\pi as\tilde{G}(s)} \tag{5}$$

where $\tilde{G}(s)$ is the Laplace transform of the stress relaxation modulus, s is the Laplace frequency, and a is the radius of the particles. An alternative expression for the GSE equation can be written in the Fourier domain as:

$$G^*(\omega) = \frac{k_BT}{\pi a i \omega \Im \left\langle \Delta r^2(t) \right\rangle} \tag{6}$$

where $\Im$ represents a unilateral Fourier transform, which is effectively a Laplace transform generalized for a complex frequency iω. Different methods have been devised to obtain $\tilde{G}(s)$ from the experimental MSD including direct Laplace or Fourier transformations (Dasgupta et al., 2002; Evans et al., 2009), or analytical approximations (Mason, 2000; Wu & Dai, 2006). It must be stressed that the GSE equation is valid under the following approximations: (a) the medium around the sphere may be treated as a continuum material, which requires that the size of the particle be larger than any structural length scale of the material, (b) no slip boundary conditions, (c) the fluid surrounding the sphere is incompressible, and (d) no inertial effects.

The application of the GSE is limited to a frequency range limited in the high frequency range by the appearance of inertial effects. The high frequency limit is imposed by the fact that the viscous penetration depth of the shear waves propagated by the particle motion must be larger than the particle size. The penetration depth is proportional to $(G*/\rho\omega^2)^{1/2}$, where ρ is the density of the fluid surrounding the particles, and for micron-sized particles in water is of the order of 1 MHz. On the other hand, the lower limit is set by the time at which compressional modes become significant compared to the shear modes excited by the particle motion. An approximate value for the low frequency limit is given by

$$\omega_L \geq \frac{G'\xi^2}{\eta a} \tag{7}$$

ξ being the characteristic length scale of the elastic network in which the particles move. Again, for the same conditions mentioned above, the low-frequency limit is in the range of 0.1 to 10 Hz. Figure 6.a shows the frequency dependence of the shear modulus for a 3D gel using two passive techniques: DWS and particle tracking. As it can be observed the agreement is very good. It must be stressed that, in order to obtain reliable Laplace or Fourier transforms of the MSD, it is necessary to measure the particle trajectories over long t periods (minutes), which makes absolutely necessary to eliminate any collective drift in the system. Very recently Felderhof (2009) has presented an alternative method for calculating the shear complex modulus from the velocity autocorrelation function, VAF, that can be calculated from the particle trajectories. An experimental difficulty associated to this method is that the VAF decays very rapidly, and therefore it is difficult to obtain many experimental data in the decay region.

Under the same conditions assumed for the GSE equation, the creep compliance is directly related to the MSD by

$$J(t) = \frac{\pi a}{k_B T} \langle \Delta r^2(t) \rangle \tag{8}$$

Even though the GSE method has been applied to different bulk systems, few applications have been done for studying the complex shear modulus of interfaces and thin films (Wu & Dai, 2006; Prasad & Weeks, 2009; Maestro et al., 2011).

The two-point correlation method also provides information about the viscoelastic moduli of the fluid in which the particles are embedded. In effect, the ensemble averaged tensor product, Eq.(4), leads to (Chen et al., 2003)

$$\tilde{D}_{rr}(r,s) = \frac{k_B T}{2\pi rs\tilde{G}(s)}; \quad D_{\theta\theta} = D_{\phi\phi} = \frac{1}{2}D_{rr} \tag{9}$$

where $\tilde{D}_{rr}(r,s)$ is the Laplace transform of $D_{rr}(r,t)$ and the off-diagonal terms vanish. Figure 6.b compares the G' and G" results calculated for a solution of F-actin (MSD data shown in Figure 4) using one- and two-particle tracking methods. The results agree with those obtained by single-particle methods as far as the scale of the inhomogeneities is similar to the particle size, otherwise the single particle method is affected both by local and global rheology. Notice that the results of the two-point technique agree with those obtained with conventional macroscopic rheometers.

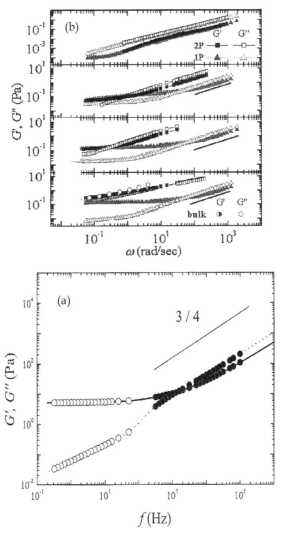

Fig. 6. Real and imaginary components calculated from the MSD shown in: a) the Figure 4.a, and b) Figure 5. Notice the good agreement between the results calculated from DWS (closed symbols) and single particle tracking (open symbols) in Figure 5.a. The solid and dotted lines are guides for G′and G″ results, respectively. In Figure 6.b the open symbols refer to G″, and the full ones to G′. Triangles correspond to single particle tracking and squares to two-particle tracking. Circles correspond to conventional macro-rheology. Figure 6.a was taken from Vincent et al. (2007) and Figure 6.b from Cherdhirankorn et al. (2009).

3. Dynamics of particles at interfaces

For using particle tracking techniques to get insight of the interfacial microrheology it is first necessary to study the diffusion of particles in the bare interface. For an inviscid interface the drag comes entirely from the upper and lower fluid phases (in the usual air-water interface only from the water subphase). The MSD of particles trapped at fluid interfaces depends on the surface concentration, and for very low surface concentration it is linear with time, thus the diffusion coefficients, D_0, can be easily obtained. However, for high surface concentrations, even below the threshold of aggregation or fluid-solid phase transitions (Bonales et al., 2011), the MSD is no longer linear with time, but shows a sub-diffusive behavior, MSD(t) ~ t^{α} with $\alpha<1$, hence D_0 must be obtained from the time dependence of the MSD in the limit of short times.

3.1 Shear micro-rheology of monolayers at fluid interfaces

In the case of particles trapped at interfaces Einstein's equation, Eq.(3), is still valid. However, one cannot calculate the friction coefficient using Stokes equation and directly substituting the interfacial shear viscosity. Instead, f is a function of the viscosities of the phases (η's), the geometry of the particle (the radius "a" for spheres), the contact angle between the probe particle and the interface (θ), etc. For a pure 2D system there is no solution for the slow viscous flow equations for steady translational motion of a sphere in a 2D fluid (Stokes paradox).

3.1.1 Motion of a disk in and incompressible membrane of arbitrary viscosity

Saffman & Delbrück (1975) and Hughes et al. (1981) have solved the problem of the motion of a thin disk immersed in a membrane of arbitrary viscosity, η_L separating two phases of viscosities η_1 and η_2. The height of the disk is assumed to be equal to the membrane thickness, h. They obtained the following expression for the translational mobility,

$$b_T = \frac{1}{f} = \frac{1}{4\pi(\eta_1 + \eta_2)R\Lambda(\varepsilon)} \tag{10}$$

Where $\Lambda(\varepsilon)$ is non-linear function of ε, $\left[\varepsilon = \frac{R}{h}\left(\frac{\eta_1 + \eta_2}{\eta_L}\right)\right]$. $\Lambda(\varepsilon)$ cannot be expressed analytically except for two limit cases,

$$\Lambda(\varepsilon) = \left\{\varepsilon\left(\ln\left(\frac{2}{\varepsilon}\right) - \gamma + \frac{4}{\pi}\varepsilon - \frac{1}{2}\varepsilon^2\ln\left(\frac{2}{\varepsilon}\right) + O(\varepsilon^2)\right)\right\}^{-1} \qquad \text{(Highly viscous membranes, e<1)}$$

$$\Lambda(\varepsilon) = \frac{2}{\pi} \qquad \text{(Low viscous membranes, } \varepsilon>1\text{)}$$

These works have been generalized by Stone & Adjari (1998) and by Barentin et al. (2000).

3.1.2 Danov's model for a sphere in a compressible surfactant layer

The above theories are limited to non protruding particles (or high membrane viscosities). In particle tracking experiments spherical particles are used that are partially immersed in both fluid phases separating the interface. Danov et al. (1995) and Fischer et al. (2006) have made numerical calculations of the drag coefficient of spherical microparticles trapped at fluid-fluid interfaces. While Danov considered the interface as compressible, Fischer assumed that the interface is incompressible, both authors predicted the dynamics of the particles adsorbed on bare fluid interfaces, i.e. with no surfactant monolayers (the so-called the limit of cero surface viscosity). The predictions of their theories are different, and will be discussed in detail below. More recently, Reynaert et al. (2007) and Madivala et al. (2009) have studied the dynamics of spherical, weakly aggregated, and of non-spherical particles at interfaces, though using macroscopic rheometers.

Danov et al. (1995) have calculated the hydrodynamic drag force and the torque acting on a micro spherical particle trapped at the air-liquid interface (they consider the viscosity of air to be zero) interface, and moving parallel to it. This model was later extended by Dimova et al. (2000) and by Danov et al. (2000) to particles adsorbed to flat or curved (spherical) interfaces separating two fluids of non vanishing viscosity. The interface was modeled as a *compressible*, 2D fluid characterized by two dimensionless parameters K and E defined as $E = \eta_{sh}/(\eta a)$ and $K = \eta_d/(\eta a)$, being η_{sh} and η_d the surface shear and dilational viscosity respectively (Note that E is the inverse of ε used by Hughes). Danov et al. made the following assumptions: 1) The movement implies a low Reynolds number, thus they ignored any inertial term; 2) the moving particle is not affected by capillarity or electro-dipping; 3) the contact line does not move to respect to the particle surface, and 4) they considered E=K, i.e. the interface is compressible. With these assumptions they solved numerically the Navier-Stokes equation to obtain the values of the drag coefficient f as a function the contact angle and of E (or K). They presented their results in graphical form, and their results are reproduced in Figure 7.

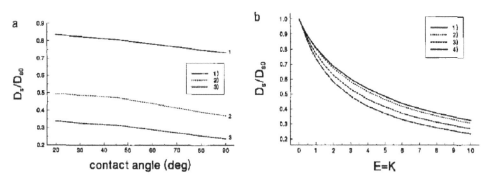

Fig. 7. Left: Effect of contact angle on the diffusion coefficient of a particle trapped at a fluid interface according to Danov's theory. D_{s0} is the diffusion coefficient for the bare interface. The different lines correspond to the following values of E (=K): 1) 0; 2) 1; 3) 5. Right: Effect of the surface to bulk shear viscosity on the diffusion coefficient. The different lines correspond to the following values of E (=K): 1) 0; 2) 1; 3) 5; 4) 10. Figures reproduced from Dimova et al. (2000).

These curves can be used to obtain the shear viscosity of compressible surfactant layer once one has obtained the diffusion coefficient from particle tracking experiments, D_0, for a free interface and in the presence of a surfactant layer. It must be stressed that, from a strict theoretical point of view, the results presented by Danov are valid only in the limit $E \gg 1$, and for arbitrary values of the contact angle. Sickert & Rondelez (2003) were the first to applied Danov's ideas to obtain the surface shear viscosity by particle tracking using spherical microparticles trapped at the air-water interface, which was covered with Langmuir films. They have measured the surface viscosity of three monolayers formed by pentadecanoic acid (PDA), L-a-dipalmitoylphosphatidylcholine (DPPC) and N-palmitoyl-6-n-penicillanic acid (PPA) respectively. The values of the shear viscosities for PDA, DPPC and PPA reported were in the range of 1 to 11.10^{-10} $N \cdot s \cdot m^{-1}$ in the liquid expanded region of the monolayer. These values are beyond the range that can be reached by macroscopic mechanical methods, that usually have a lower limit in the range of 10^{-7} $N \cdot s \cdot m^{-1}$.

Fischer considered that a monolayer cannot be considered as compressible. Due to the presence of a surfactant, Marangoni forces (forces due to surface tension gradients) strongly suppress any motion at the surface that compress or expands the interface. Any gradient in the surface pressure is almost instantly compensated by the fast movement of the surfactant at the interface given a constant surface pressure, behaving thus as a incompressible monolayer (Fischer assumed that the velocity of the 2D surfactant diffusion is faster than the movement of the beads). The fact that the drag on a disk in a monolayer is that of an incompressible surface has been verified experimentally by Fischer (2004). In the case of Langmuir films of polymers, the monolayer could be considered as compressible or incompressible depending on the rate of the polymer dynamics at the interface compared to the velocity of the beads probes. Bonales et al. (2007) have calculated the shear viscosity of two polymer Langmuir films using Danov's theory, and compared these values with those obtained by canal viscosimetry. Video Particle tracking and Danov's theory were used by Maestro et al. (2011.a) to show the glass transition in Langmuir films. Figure 8 shows the results obtained for a monolayer of poly(4-hydroxystyrene) onto water. For all the monolayers reported by Bonales et al. (2007) and Maestro et al. (2011.b) the surface shear viscosity calculated from Danov's theory was lower than that measured with the macroscopic canal surface viscometer. Similar qualitative conclusions were reached at by Sickert et al. (2007) for monolayers of fatty acids and phospholipids in the liquid expanded region.

3.1.3 Fischer's theory for a sphere in a incompressible surfactant layer

Fischer et al. (2006) have numerically solved the problem of a sphere trapped at an interface with a contact angle θ moving in an *incompressible* surface. They showed that contributions due to Marangoni forces account for a significant part of the total drag. This effect becomes most pronounced in the limit of vanishing surface compressibility. In this limit the Marangoni effects are simply incorporated to the model by approximating the surface as incompressible. They solved the fluid dynamics equations for a 3D object moving in a monolayer of surface shear viscosity, η_s between two infinite viscous phases. The monolayer surface is assumed to be flat (no electrocapillary effects). Then the translational drag coefficient, $k_{T,}$ was expressed as a series expansion of the Boussinesq number, $B = \eta_s / ((\eta_1 + \eta_2) \cdot a)$, a being the radius of the spherical particle:

$$k_T = k_T^0 + B k_T^1 + O(B^2) \tag{11}$$

For B=0, and for an air-water interface (η_1, η_2=0), the numerical results for k_T are fitted with an accuracy of 3% by the formula,

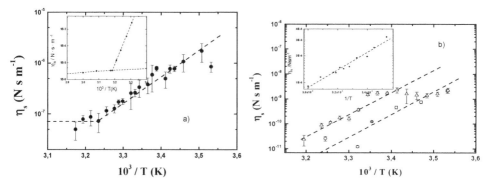

Fig. 8. Temperature dependence of the surface shear viscosity of a monolayer of poly(4-hydroxystyrene) at the air-water interface obtained by particle tracking (the insets show the corresponding values measured with a macroscopic canal viscometer. Left: experiments done at Π=8 mN·m^{-1}. Right: triangles correspond to Π=3 mN·m^{-1} and circles to Π=2 mN·m^{-1}. Notice that the results obtained by particle tracking are much smaller than those obtained with the canal viscometer. Data taken from Hilles et al. (2009).

$$k_T^0 \approx 6\pi \sqrt{\tanh\left(32\left(\frac{d}{R} + 2 \right) \middle/ \left(9\pi^2 \right) \right)} \qquad (12)$$

where d is the distance from the apex of the bead to the plane of the interface (which defines the contact angle). Note that if d goes to infinity, $k_T^0 = 6\pi$, which is the correct theoretical value for a sphere in bulk (Stokes law). The translational drag in a half immersed sphere in a non viscous monolayer is $k_T^0 \approx 11$ which is about 25% higher than the drag on a sphere trapped at a free surface, $k_T = 3\pi$. This means that even in the absence of any appreciable surface viscosity the drag coefficient of an incompressible monolayer is higher than that of a free interface, and the data cannot be used to extract the surface shear viscosity using Danov's theory especially in the limit of low surface viscosities.

The numerical results for $k_T^{(1)}$ are fitted within an accuracy of 3% to,

$$k_T^{(1)} \approx \begin{cases} -4\ln\left(\frac{2}{\pi}\arctan\left(\frac{2}{3} \right) \right)\left(\frac{a^{3/2}}{(d+3a)^{3/2}} \right) & (d/a>0) \\ -4\ln\left(\frac{2}{\pi}\arctan\left(\frac{d+2a}{3a} \right) \right) & (d/a<0) \end{cases} \qquad (13)$$

Sickert & Rondelez (2003) have introduced in an ad-hoc way the incompressibility effect in Danov's theory by renormalizing his master curve (Figure 7 above) by the empirical value of 1.2, and they have later reanalyzed their data by combining the Danov's and Fischer's theories (Sickert et al., 2007). First they used the value determined by Danov et al. (2000) for the resistance coefficient of a sphere at a clean, compressible surface and at the contact angle of their experiments (50°). Afterwards, they used the predictions of Fischer et al. (2006) for a sphere in a surfactant monolayer (incompressible) with the contact angle corrected by the change in the surface tension, and in the case of E <<<1 (notice that this is the opposite E-limit than for the original Danov's theory),

$$\frac{D_0}{D_{\to 0}} = \frac{\xi_0^{(\text{Danov})}(\theta)}{\xi^{(\text{Fisher})}(\theta)} = \frac{\xi_0^{(\text{Danov})}(\theta)}{k_T^0(\theta) + Ek_T^1(\theta)} \qquad (14)$$

D_0 being the diffusion coefficient of the beads at a free surface (compressible), and $D_{\to 0}$ is the value of an incompressible monolayer which surface concentration is tending to zero. They found that this relation is not equal to 1 but to 0.84 for their systems and experimental conditions which confirms the observation of Barentin et al. (2000).

Figure 9 shows the friction coefficient for latex particles at the water-air interface obtained from particle tracking for polystyrene latex particles. It also shows the values calculated from Danov's and from Fischer's theories (notice that for the bare interface E = B =0). The figure clearly shows that both theories underestimate the experimental values over the whole θ range. An empirical factor of $\eta(\theta)_{\text{exp}}/\eta(\theta)_{\text{Fisher}} = 1.8\pm0.2$ brings the calculated values in good agreement with the experiments at all the contact angle values. A similar situation was found for the water-n-octane interface.

The values of the shear viscosities calculated by Sickert & Rondelez (2003) by using the modified-Fisher theory are 2 or 3 times higher than the previous values. Sickert et al. (2007) also refers to a model developed by Stone which would be valid over the whole range of E, although only for a contact angle of 90°. Figure 10 shows clearly the large difference found between micro- and macrorheology for monolayers of poly(t-butyl acrylate) at the so-called Γ** surface concentration (Muñoz et al., 2000). The macrorheology results have been obtained using two different oscillatory rheometers. The huge difference cannot be attributed to specific interactions between the particles and the monolayer.

In effect, Figure 11 shows that the values obtained are the same for particles of rather different surface characteristics. Moreover, the values calculated from the modified-Fisher's theory or by direct application of the GSE equation lead to almost indistinguishable surface shear viscosities. It must be stressed that in all the cases the contact angle used is the experimentally measured using the gel-trapping technique described by Paunov et al.

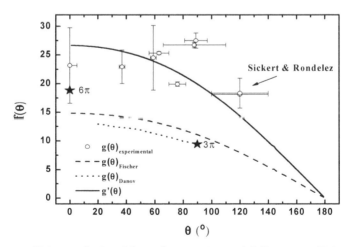

Fig. 9. Friction coefficients calculated from the experimental diffusion coefficients measured by particle tracking experiments (symbols), by Danov's theory (dotted line), by Fischer's theory (dashed line), and by the corrected Fischer's theory (continuous line).

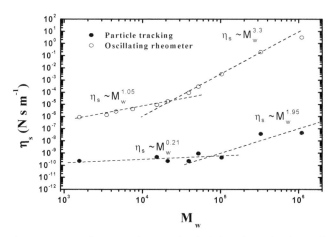

Fig. 10. Surface shear viscosity for monolayers of poly(t-butyl acrylate) as a function of the molecular weight and for a surface pressure of 16 mN·m^{-1}. The lower curve corresponds to data obtained from particle tracking. The upper curve was obtained from conventional oscillatory rheometers.

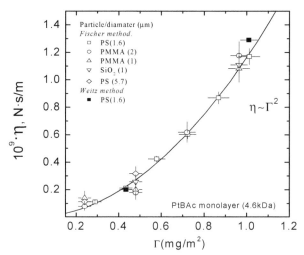

Fig. 11. Surface shear viscosity of a monolayer of poly(t-butyl acrylate) (molecular weight 4.6 kDa) measured by particle tracking. Different microparticles where used: poly(styrene) of 1.6 and 5.7 μm (stabilized by sulfonate groups); poly(methylmethacrylate) stabilized by Coulombic repulsions (PMMA1), or by steric repulsions (PMMA2); Silica particles stabilized by Coulombic repulsions. Empty symbols: the viscosities were calculated using Fischer theory. Full symbols: calculated by the GSE equation.

(2003). This discrepancy between micro- and macrorheology in the study of monolayers seems to be a rather general situation (Schmidt et al., 2000; Khair & Brady, 2005; Oppong & de Bruyn, 2010; Lee et al., 2010) and no clear theoretical answer has been found so far.

4. Conclusions

The set of microrheological techniques offer the possibility of studying the rheology of very small samples, of systems which are heterogeneous, and facilitate to measure the shear modulus over a broad frequency range. Particle tracking techniques are especially well suited for the study of the diffusion of microparticles either in the bulk or at fluid interfaces. Different types of mean squared displacements, MSD, (one-particle, two-particle) allow one to detect spatial heterogeneities in the samples. Even though good agreement has been found between micro- and macrorheology (at least when two-particle MSD is used) in bulk systems, the situation is still not clear for the case of fluid interfaces, where the shear surface microviscosity is much smaller than the one measured with conventional surface rheometers.

5. Acknowledgments

This work has been supported in part by MICIN under grant FIS2009-14008-C02-01, by ESA under grant FASES MAP-AO-00-052, and by U.E. under grant Marie-Curie-ITN "MULTIFLOW". L.J. Bonales and A. Maestro are grateful to MICINN for their Ph.D. fellowships. We are grateful to Th.M. Fisher, R. Miller and L. Liggieri for helpful discussions.

6. References

Alexander M., Dalgleish D.G., 2007, Diffusing wave spectroscopy of aggregating and gelling systems. *Curr. Opinion Colloid Interf. Sci.* 12 (4-5) 179-186, ISSN: 1359-0294

Barentin C., Muller, P., Ybert, C., Joanny J.-F., di Meglio J.-M., 2000, Shear viscosity of polymer and surfactant monolayers. *Eur. Phys. J. E.* 2 (2) 153-159, ISSN: 1292-8941

Bausch A.R., Hellerer U., Essler M., Aepfelbacher M., Sackmann E., 2001, Rapid stiffening of integrin receptor-actin linkages in endothelial cells stimulated with thrombin: A magnetic bead microrheometry study. *Biophys. J.* 80 (6) 2649-2657, ISSN: 0006-3495

Binks B., Horozov S., (Eds.), 2006, *Colloidal particles at liquid interfaces.* Cambridge Univ. Press, ISBN: 9780521848466, Cambridge.

Bonales L.J., Ritacco H., Rubio J.E.F., Rubio R.G., Monroy F., Ortega F., 2007. Dynamics in ultrathin films: Particle tracking microrheology of Langmuir monolayers. *Open Phys. Chem. J.* 1 (1) 25-32; ISSN: 1874-0677

Bonales L.J., Rubio J.E.F., Ritacco H., Vega C., Rubio R.G., Ortega F., 2011. Freezing transition and interaction potential in monolayers of microparticles at fluid interfaces. *Langmuir* 27 (7) 3391-3400, ISSN: 0743-7463.

Borsali R., Pecora R., (Eds.), 2008. *Soft-matter characterization.* Vol. 1., Springer, ISBN: 978-1-4020-8290-0, Berlin.

Breedveld V., Pine D.J., 2003, Microrheology as a tool for high-throughput screening. *J. Mater. Sci.* 38 (22), 4461-4470, ISSN: 002-2461

Chen D.T., Weeks E.R., Crocker J.C., Islam M.F., Verma R., Gruber J., Levine A.J., Lubensky T.C., Yodh A.G., 2003, Rheological microscopy. Local mechanical properties from microrheology. *Phys. Rev. Lett.* 90 (10) 108301, ISSN: 0031-9007

Cherdhirankorn T., Harmandaris V., Juhari A., Voudouris P., Fytas G., Kremer K., Koynov
     K., 2009, Fluorescence correlation spectroscopy study of molecular probe diffusion
     in polymer melts. *Macromolecules* 42 (13) 4858-4866, ISSN: 0024-9297
Cicuta P., Donald A.M., 2007, Microrheology: a review of the method and applications. *Soft
     Matter* 3 (12) 1449-1455, ISSN: 1744-683X
Cicuta P., Keller S.L., Veatch S.L., 2007, Diffusion of Liquid Domains in Lipid Bilayer
     Membranes , *J. Phys. Chem. B* 111 (13) 3328-3331, ISSN:1520-6106
Conroy   R., 2008, Force spectroscopy with optical and magnetic tweezers. In *Handbook of
     molecular force spectroscopy*. Noy A. (Ed.), Springer, ISBN: 978-0-387-49987-1, Berlin.
Crocker J.C., Grier D.G., 1996, Methods of digital video microscopy for colloidal studies. *J.
     Colloid Interf. Sci.* 179 (1), 298-310, ISSN: 0021-9797
Danov K., Aust R., Durst F., Lange U. 1995. Influence of the surface viscosity on the
     hydrodynamic resistance and surface diffusivity of a large Brownian particle. *J.
     Colloid Int. Sci.*, 175 (1), 36-45, ISSN: 0021-9797
Danov K.D., Dimova R., Pouligny B. 2000. Viscous drag of a solid sphere stradding a
     spherical or flat surface. *Physics of Fluids*, 12 (11) 2711-2722, ISSN: 0899-8213
Dasgupta B.R., Tee S.Y., Crocker J.C., Frisken B.J., Weitz D.A., 2002, Microrheology of
     polyethylene oxide using diffusing wave spectroscopy and single scattering. *Phys.
     Rev. E* 65 (5) 051505, ISSN: 1539-3755
Dasgupta B.R., Weitz D.A., 2005, Microrheology of cross-linked polyacrylamide networks.
     Phys. Rev. E 71 (2) 021504, ISSN: 1539-3755
Díez-Pascual A.M., Monroy F., Ortega F., Rubio R.G., Miller R., Noskov B.A., 2007,
     Adsorption of water-soluble polymers with surfactant character. Dilational
     viscoelasticity. *Langmuir* 23 (7) 3802-3808, ISSN: 0743-7463
Dimova R., Danov K., Pouligny B., Ivanov I.B. 2000. Drag of a solid particle trapped in a thin
     film or at an interface: Influence of surface viscosity and elasticity. *J. Colloid Interf.
     Sci.* 226 (1) 35-43, ISSN: 0021-9797
Evans R.M., Tassieri M., Auhl D., Waigh Th.A., 2009, Direct conversion of rheological
     compliance measurements into storage and loss moduli. *Phys. Rev. E* 80 (1) 012501,
     ISSN: 1539-3755
Fabry B., Maksym G.N., Butler J.P., Glogauer M., Navajas D., Fredberg JJ. 2001, Scaling the
     microrheology of living cells. *Phys. Rev. Lett.* 87 (14) 148102, ISSN: 0031-9007
Feder T.J., Brust-Mascher I., Slattery J.P., Barid B., Webb W.W., 1996, Constrained diffusion
     or immobile fraction on cell surfaces: A new interpretation. *Biophys. J.* 70 (6) 2767-
     2773, ISSN: 0006-3495
Felderhof B.U., 2009, Estimating the viscoelastic moduli of a complex fluid from observation
     of Brownian motion. *J. Chem. Phys.* 131 (16) 164904, ISSN: 0021-9606
Fischer Th.M. 2004. Comment on "Shear viscosity of Langmuir Monolayers in the Low
     Density Limit". *Phys. Rev. Lett.* 92 (13) 139603, ISSN: 0031-9007
Fischer Th. M., Dhar P., Heinig P. 2006. The viscous drag of spheres and filaments moving in
     membranes or monolayers, *J. Fluid Mech.* 558 (1) 451-475, ISSN: 0022-1120
Gardel M.L., Valentine M.T., Weitz D.A., 2005, Microrheology. In *Microscale diagnostic
     techniques*. Brauer K., ed. Springer, ISBN: 978-3-540-23099-1, Berlin.

Gavranovic G.T., Deutsch J.M., Fuller G.G., 2005, Two-dimensional melts: Chains at the air-water interface. *Macromolecules*, 38 (15) 6672-6679, ISSN: 0024-9297

Guo J., Baker G.A., Hillesheim P.C., Dai S., Shaw R.W., Mahurin S.M. 2011. Fluorescence correlation spectroscopy evidence for structural heterogeneity in ionic liquids. *Phys. Chem. Chem. Phys.* 13 (27), 12395-12398, ISSN: 1463-9076

Hasnain I., Donald A.M., 2006, Microrheology characterization of anisotropic materials. *Phys. Rev. E* 73 (3) 031901, ISSN: 1539-3755

Heuf R.F., Swift J.L., Cramb D.T., 2007, Fluorescence correlation spectroscopy using quantum dots: Advances, challenges and opportunities. *Phys. Chem. Chem. Phys.* 9 (16) 1870-1880, ISSN: 1463-9076

Hilles H.M., Ritacco H., Monroy F., Ortega F., Rubio R.G. 2009. Temperature and concentration effects on the equilibrium and dynamic behavior of a Langmuir monolayer: From fluid to gel-like behavior. *Langmuir* 25 (19) 11528-11532, ISSN: 0743-7463

Hughes B.D., Pailthorpe P.A., White L.P. 1981. The translational and rotational drag of a cylinder moving in a membrane. *J. Fluid Mechanics* 110 (1) 349-372, ISSN: 0022-1120

Kahir, A.S., Brady, J.F., 2005, "Microviscoelasticity" of colloidal dispersions, *J. Rheol.*, 49 (6) 1449-1481; ISSN: 0148-6055

Keller M., Schilling J., Sackmann E., 2001, Oscillatory magnetic bead rheometer for complex fluid microrheometry. *Rev. Sci. Instrum.* 72 (9) 3626-3634, ISSN: 0034-6748

Konopka M.C., Weisshaar J.C., 2004, Heterogeneous motion of secretory vesicles in the actin cortex of live cells: 3D tracking to 5-nm accuracy. *J. Phys. Chem. A* 108 (45) 9814-9826, ISSN: 1520-6106

Kralchewski P., Nagayama K., 2001, Particles at fluid interfaces, attachment of colloid particles and proteins to interfaces and formation of two-dimensional arrays. In *Studies in interfacial science*, Vol. 10., Möbius D., Miller R., (Eds.), Elsevier, ISBN: 978-0-444-52180-4, Amsterdam.

Langevin D., 2000, Influence of interfacial rheology on foam and emulsion properties. *Adv. Colloid Interf. Sci.* 88 (1-2) 209-222, ISSN: 0001-8686

Larson R.G., 1999, *The structure and rheology of complex fluids*. Oxford University Press, ISBN: 978-0195121971, New York.

Lee M.H., Lapointe C.P., Reich D.H., Steebe K.J., Leheny R., 2009, Interfacial hydrodynamic drag on nanowires embedded in thin oil films and protein layers, *Langmuir* 25 (14) 7976-7982, ISSN: 0743-7463

Lee M.H., Reich D.H., Steebe K.J., Leheny R.L., 2010, Combined passive and active microrheology study of protein-layer formation at an air-water interface, *Langmuir* 26 (4) 2650-2658, ISSN: 0743-7463

Levine A.J., Lubensky T.C., 2000, One- and two-particle microrheology. *Phys. Rev. Lett.* 85 (8) 1774-1777, ISSN: 0031-9007

Lin B., Rice S.A., Weitz D.A., 1993, Static and dynamic evanescent wave light scattering studies of diblock copolymers adsorbed at the air/water interface. *J. Chem. Phys.*, 99 (10) 8308-8324, ISSN: 0021-9606

Liu J., Gardel M.L., Kroy K., Frey E., Hoffman B.D., Crocker J.C., Bausch A.R., Weitz D.A., 2006, Microrheology probes length scale dependent rheology. *Phys. Rev. Lett.* 96 (11) 118104, ISSN: 0031-9007

MacKintosh F.C., Schmidt C.F., 1999, Microrheology. *Curr. Opinion Colloid Interf. Sci.* 4 (4), 300-307, ISSN: 1359-0294

Madivala B., Fransaer J., Vermant J. 2009. Self-assembly and rheology of ellipsoidal particles at interfaces. *Langmuir* 25 (5) 2718-2728, ISSN: 0743-7463

Maestro A., Guzmán E., Chuliá R., Ortega F., Rubio R.G., Miller R. 2011.a. Fluid to soft-glass transition in a quasi-2D system: Thermodynamic and rheological evidences for a Langmuir monolayer. *Phys. Chem. Chem. Phys.* 13 (20) 9534-9539, ISSN: 1463-9076

Maestro A., Bonales L.J., Ritacco H., Fischer Th.M., Rubio R.G., Ortega, F., 2011.b. Surface rheology: Macro- and microrheology of poly(tert-butyl acrylate) monolayers. *Soft Matter*, in press,ISSN: 1744-683X

Marcus A.H., Lin B., Rice S.A., 1996, Self-diffusion in dilute quasi-two-dimensional hard sphere suspensions: Evanescent wave light scattering and video microscopy studies. *Phys. Rev. E* 53 (2) 1765-1776, ISSN: 1539-3755

Mason T.G., Weitz D.A., 1995, Optical measurements of frequency-dependent linear viscoelastic moduli of complex fluids. *Phys. Rev. Lett.* 74 (7) 1250-1253, ISSN: 0031-9007

Mason Th.G., 2000, Estimating the viscoelastic moduli of complex fluids using the generalized Stokes-Einstein equation. *Rheol. Acta*, 39 (4) 371-378, ISSN: 0035-4511

Miller R., Liggieri L., (Eds.), 2009, *Interfacial rheology*. Brill, ISBN: 9789004175860, Leiden.

Mukhopadhyay A., Granick S., 2001, Micro- and nanorheology, *Curr. Opinion Colloid Interf. Sci.* 6 (5-6),423-429, ISSN: 1359-0294

Muñoz M.G., Monroy, F., Ortega F., Rubio R.G., Langevin D., 2000, Monolayers of symmetric triblock copolymers at the air-water interface. 2. Adsorption kinetics. *Langmuir* 16 (3) 1094-1101, ISSN: 0743-7463

Oppong F.K., de Bruyn J.R., 2010, Microrheology and dynamics of an associative polymer, *Eur. Phys. J. E* 31 (1) 25-35, ISSN: 1292-8941

Ou-Yang H.D., Wei M.T., 2010, Complex fluids: Probing mechanical properties of biological systems with optical tweezers, *Ann. Rev. Phys. Chem.*, 61, 421-440, ISSN: 0066-426X

Pan W., Filobelo L., Phan N.D.Q., Galkin O., Uzunova V.V., Vekilov P.G., 2009, Viscoelasticity in homogeneous protein solutions, *Phys. Rev. Lett.* 102 (5) 058101, ISSN: 0031-9007

Paunov V.N. 2003. Novel Method for Determining the Three-Phase Contact Angle of Colloid Particles Adsorbed at Air–Water and Oil–Water Interfaces. *Langmuir* 19 (19) 7970–7976, ISSN: 0743-7463

Prasad V., Weeks E.R., 2009, Two-dimensional to three-dimensional transition in soap films demonstrated by microrheology, *Phys. Rev. Lett.* 102 (17) 178302, ISSN: 0031-9007

Resnick A., 2003, Use of optical tweezers for colloid science, *J. Colloid Interf. Sci.* 262 (1) 55-59, ISSN: 0021-9797

Reynaert S., Moldenaers P., Vermant J. 2007. Interfacial rheology of stable and weakly aggregated two-dimensional suspensions. *Phys. Chem. Chem. Phys.* 9 (48) 6463-6475, ISSN: 1463-9076

Reynaert S., Brooks C.F., Moldanaers P., Vermant J., Fuller G.G., 2008. Analysis of the magnetic rod interfacial stress rheometer. 52 (1), 261-285, ISSN: 0148-6055

Riande E., Díaz-Calleja R., Prolongo M.G., Masegosa R.M., Salom C., 2000, *Polymer viscoelasticity. Stress and strain in practice.* Marcel Dekker, ISBN: 0-8247-7904-5, New York.

Ries J., Schwille P., 2008, New concepts for fluorescence correlation spectroscopy on membranes. *Phys. Chem. Chem. Phys.* 10 (24) 3487-3497, ISSN: 1463-9076

Riegler R., Elson E.S., 2001, *Fluorescence correlation spectroscopy: Theory and applications.* Springer-Verlag, ISBN; 978-3540674337, Berlin.

Saffman P-G., Delbrück M. 1975. Brownian Motion in Biological Membranes. *Proc. Nat. Acad. Sci. USA* 72 (8) 3111-3113, ISSN: 1091-6490

Saxton M.J., Jacobson K., 1997, Single-particle tracking: Applications to membrane dynamics. *Annu. Rev. Biophys. Biomol Struct.* 26 (1) 373-399, ISSN: 1056-8700

Schmidt F.G., Hinner B., Sackmann E., 2000, Microrheometry underestimates the values of the viscoelastic moduli in measurements on F-actin solutions compared to macrorheometry, *Phys. Rev. E* 61 (5) 5646-5652, ISSN: 1539-3755

Sickert M., Rondelez F. 2003. Shear viscosity of Langmuir monolayers in the low density limit. *Phys. Rev. Lett.* 90 (12) 126104, ISSN: 0031-9007

Sickert M., Rondelez F., Stone H.A. 2007. Single-particle Brownian dynamics for characterizing the rheology of fluid Langmuir monolayers. *Eur. Phys. Lett.* 79 (6) 66005, ISSN: 0295-5075

Squires T.M., 2008, Nonlinear microrheology: Bulk stresses versus direct interactions. *Langmuir* 24 (4) 1147-1159, ISSN: 0743-7463

Squires T.M., Mason Th.G., 2010, Fluid mechanics of microrheology, *Ann. Rev. Fluid Mechanics*, 42 (1) 413-438, ISSN: 0066-4189

Steffen P., Heinig P., Wurlitzer S., Khattari Z., Fischer Th.M., 2001, The translational and rotational drag on Langmuir monolayer domains, *J. Chem. Phys.* 115 (2) 994-997, ISSN: 0021-9606

Stone H., Adjari A. 1998. Hydrodynamics of particles embedded in a flat surfactant layer overlying a subphase of finite depth. *J. Fluid Mech.* 369 (1) 151-173, ISSN: 0022-1120

Tassieri M., Gibson G.M., Evans R.M.L., Yao A.M., Warren R., Padgett M.J., Cooper J.M., 2010, Optical tweezers. Broadband microrheology, *Phys. Rev. E* 81 (2) 026308, ISSN: 1539-3755

Vincent R R , Pinder D N , Hemar Y., Williams M.A.K., 2007, Microrheological studies reveal semiflexible networks in gels of a ubiquitous cell wall polysaccharide. *Phys. Rev. E* 76 (3) 031909, ISSN: 1539-3755

Waigh T.A., 2005, Microrheology of complex fluids. *Rep. Prog. Phys.* 68 (3) 685-742, ISSN: 0034-4885

Wilson L.G., Poon W.C.K., 2011, Small-world rheology: An introduction to probe-based active microrheology, *Phys. Chem. Chem. Phys.* 13 (22) 10617-10630, ISSN: 1463-9076

Winkler R.G., 2007, Diffusion and segmental dynamics of rodlike molecules by fluorescence correlation spectroscopy. *J. Chem. Phys.* 127 (5) 054904, ISSN: 0021-9606

Wöll D., Braeken E., Deres A., de Schryver F.C., Uji-I H., Hofkens J., 2009. Polymers and single molecule fluorescence spectroscopy, what can we learn? *Chem. Soc. Rev.* 38 (2), 313-328, ISSN: 0306-0012

Wu J., Dai L.L., 2006, One-particle microrheology at liquid-liquid interfaces. *Appl. Phys. Lett.,* 89 (9) 094107, ISSN: 0003-6951

Yoon Y.-Z., Kotar J., Yoon G., Cicuta P., 2008, The nonlinear mechanical response of the red blood cell. *Phys. Biol.* 5 (3) 036007, ISSN: 1478-3975

10

Hydrodynamic Properties of Aggregates with Complex Structure

Lech Gmachowski
Warsaw University of Technology
Poland

1. Introduction

Hydrodynamic properties of fractal aggregates and polymer coils, such as sedimentation velocity, permeability, translational and rotational diffusion coefficients and intrinsic viscosity, are of great interest in hydrodynamics, engineering, colloid and polymer science and biophysics. The hydrodynamic properties of aggregates are closely connected to their structure.

Aggregates – the clusters of monomers - usually have a fractal structure which means that parts of the object are similar to the whole. The self-similar structure is characterized by the fractal dimension which is a measure of how the aggregate fills the space it occupies. The fractal dimension can be calculated by analyzing the mass-radius relation for a series of similar aggregates, since the mass of an aggregate scales as a power of the size.

The fractal dimension can be also determined by covering the aggregate with spheres of changing radius (Fig. 1). Then plotting the number of spheres $N(\rho)$ versus their radius ρ in a log-log coordinate system, one determines the fractal dimension as the negative slope of the obtained line.

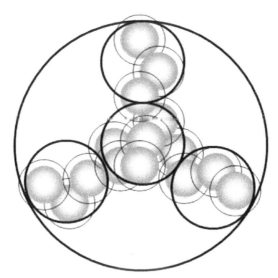

Fig. 1. Aggregate covering with spheres of changing radius.

The Hausdorff dimension (Hausdorff, 1919) is the critical exponent for which the Hausdorff measure M_d, being proportional to the product of number of spheres and a power of their radius, changes from zero to infinity when the size of covering elements tends to zero

$$M_d \propto N(\rho) \cdot \rho^d \xrightarrow{\rho \to 0} \begin{cases} 0, & d > D \\ \infty, & d < D \end{cases} \tag{1}$$

In practice each monomer has its size. It is thus generally accepted the constancy of Hausdorff measure in a finite range of size to be sufficient to characterize the aggregate structure. The constancy of the Hausdorff measure for two limiting sizes of spheres, written for an aggregate containing i monomers, can be expressed as

$$1 \cdot R^D = i \cdot R_1^D \tag{2}$$

where R is the radius of the sphere circumscribed on the aggregate and R_1 is the radius of envelope surrounding one monomer, for which the similarity to the aggregate still exists (Gmachowski, 2002).

The structure of aggregates is permeable which means that a fluid flows not only around but also through the aggregate. It is analyzed by taking into account the internal permeability of aggregates, either directly or by replacing a given aggregate by a smaller impermeable sphere of the same hydrodynamic properties. In this way the hydrodynamic radius is defined.

The structure of fractal aggregate can be related to the possibility to penetrate its interior by a fluid, well represented by internal permeability. The translational friction coefficient of a particle of radius R can be written in the following form

$$f_T = 6\pi\eta_0 R \cdot \frac{r}{R} \tag{3}$$

where the hydrodynamic radius r is introduced to take into account its dependence on the internal permeability of the aggregate. Such relation gave Brinkman for translational friction factor (Brinkman, 1947)

$$\frac{r}{R} = \frac{1 - \dfrac{\tanh\sigma}{\sigma}}{1 + \dfrac{3}{2\sigma^2}\left(1 - \dfrac{\tanh\sigma}{\sigma}\right)} \tag{4}$$

where $\sigma = R / \sqrt{k}$ is the reciprocal square root of dimensionless internal permeability of a sphere of uniform structure modeling the fractal aggregate. The analogous relations of the normalized hydrodynamic radius for the rotational friction coefficient and the intrinsic viscosity are slightly different, but all the three give the results which are very close to one another (Gmachowski, 2003).

For a homogeneous porous medium, being an arrangement of monosized particles, the permeability is proportional to the square of the characteristic pore size (Dullien, 1979) which is closely correlated to the size of constituents. In the case of fractal aggregate, which is not homogeneous, the fluid flow occurs mainly in the large pores. Hence their size determines the aggregate permeability.

For similar aggregates the size of large pores scales as the size of the whole aggregate. Therefore the ratio of the internal permeability and the square of aggregate radius is expected to be constant for aggregates of the same fractal dimension and to decrease with increasing fractal dimension due to the increment of the aggregate compactness (Gmachowski, 1999; Woodfield & Bickert, 2001; Bushell et al., 2002). This means that $\sigma = R / \sqrt{k}$ is a unique function of the fractal dimension of an aggregate and hence the ratio r/R is determined by D (Eq. 4). A formula

$$\frac{r}{R} = \sqrt{1.56 - \left(1.728 - \frac{D}{2}\right)^2} - 0.228 \tag{5}$$

has been derived from analysis of permeability of aggregated system (Gmachowski, 2000) and is confirmed by different hydrodynamic properties of fractal aggregates (Gmachowski, 2003).

This means that the hydrodynamic radius is proportional to aggregate radius for a given fractal dimension. The covering can be thus performed not only in the range of radii, but also in the range of hydrodynamic radii. The hydrodynamic radius of a solid monomer is its geometrical radius. For fractal aggregate the mass-hydrodynamic radius relation has the following form

$$i = \left(\frac{r}{a}\right)^D \tag{6}$$

since the hydrodynamic radius r converges to the primary particle radius a for the number of constituent particles equal to unity (Gmachowski, 2008). The mass-radius relation reads

$$i = \left(\frac{r}{R}\right)^D \left(\frac{R}{a}\right)^D \tag{7}$$

which reduces to the previous one if the aggregation number is related to the hydrodynamic radius instead to the radius. The full form of mass-radius relation has the form

$$i = \left[\sqrt{1.56 - \left(1.728 - \frac{D}{2}\right)^2} - 0.228\right]^D \left(\frac{R}{a}\right)^D \tag{8}$$

Plotting in a log-log system the aggregation number against radius for several similar aggregates, one can determine a best fit straight line whose slope is the fractal dimension and the location makes it possible to determine the monomer radius.

If the aggregate is composed of smaller aggregates instead of solid monomers, their number is correlated to the hydrodynamic radius of smaller aggregates according to the mass-radius relation of the form similar to Eq. (8).

2. Aggregates with complex structure

An aggregate has a complex structure if it consists of smaller aggregates instead of solid monomers (Fig. 2) and their fractal dimension is different from that of the whole aggregate.

In opposite, the constancy of Hausdorff measure would take place in the range of the whole aggregate hydrodynamic size down to the solid monomer size. An aggregate with complex structure is termed as aggregate with mixed statistics, since it has different fractal dimensions on different length scales. The constituent aggregates are known as blobs.

The knowledge of the hydrodynamic radius in relation to the radius of fractal aggregate of a given fractal dimension, utilized for blobs, makes it possible to replace the blobs by their hydrodynamic equivalents. In this way an aggregate with mixed statistics is reduced to fractal aggregate with the Hausdorff measure constant in the range of the whole aggregate hydrodynamic size down to the hydrodynamic size of blobs.

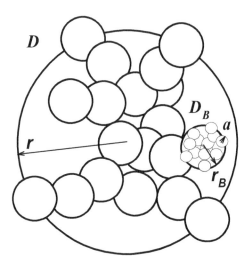

Fig. 2. Graphical representation of an aggregate with mixed statistics. The aggregate fractal dimension is a result of the spatial arrangement of blobs.

An aggregate with mixed statistics of hydrodynamic radius r and fractal dimension D consists of I blobs of hydrodynamic radius r_B and fractal dimension D_B, each containing i_B solid monomers of radius a. The mass-hydrodynamic radius relations are

$$i_B = \left(\frac{r_B}{a}\right)^{D_B} \tag{9}$$

$$I = \left(\frac{r}{r_B}\right)^{D} \tag{10}$$

Let us imagine an aggregate of the same mass and fractal dimension composed of monomers instead of blobs. Then the total number of monomers can be expressed as

$$Ii_B = \left(\frac{r_0}{a}\right)^{D} \tag{11}$$

Combining the last three equations, one gets the expression for the change of hydrodynamic radius caused by the presence of blobs

$$\frac{r}{r_0} = i_B^{1/D_B - 1/D} \tag{12}$$

The corresponding mass-radius relation for an aggregate with mixed statistics reads

$$I = \left[\sqrt{1.56 - \left(1.728 - \frac{D}{2}\right)^2} - 0.228 \right]^D \left(\frac{R}{r_B}\right)^D \tag{13}$$

which makes it possible to determine the blob hydrodynamic radius by plotting in a log-log system the number of blobs against the aggregate radius for several similar aggregates with mixed statistics and then deducing the slope and location of the best fit straight line obtained.

3. Asphaltene aggregates

The aggregate of mixed statistics can be obtained by shearing the crude oil (Gmachowski & Paczuski, 2011). Asphaltenes, a part of petroleum, are aromatic multicyclic molecules surrounded and linked by aliphatic chains and heteroatoms, of the molar mass in the range 500-50000 u. As a result of shearing, they aggregate to form blobs of a size of several micrometers, which join to form aggregates with mixed statistics. If the crude oil is mixed with toluene and n-heptane in different proportions, the range of aggregate size becomes wider.

50 μm

Fig. 3. Typical microscope image of asphaltene aggregate.

It is possible to estimate the size and number of blobs for several images (Fig. 3) to identify the form of mass-radius relation of asphaltene aggregates by plotting the data in a log-log system. This is presented in Fig. 4.

The fractal dimension determined by this method for aggregates of mixed statistics investigated was $D=1.5$, whereas the hydrodynamic radius of blobs $r_B \cong 3\mu m$. Two additional line are drawn in Fig. 4, representing Eq. (13) for the same fractal dimension and two values of the blob hydrodynamic radius, namely $r_B = 2\mu m$ and $r_B = 4\mu m$. Their locations do not correspond to points representing the experimental data, which confirms the rationality of the method of mass-radius relation for aggregates with mixed statistics.

Moreover, the size estimated (blob hydrodynamic radius $r_B = 3\mu m$) is close to that observed in the image (blob radius), which suggests very compact structure of blobs formed by asphaltenes.

4. Free settling velocity

Free settling velocity of an aggregate with mixed statistics can be determined by equating the gravitational force allowing for the buoyancy of the surrounding fluid with the opposing hydrodynamic force which depends on the aggregate size and its permeability. The use of hydrodynamic radius which is the radius of an impermeable sphere of the same mass

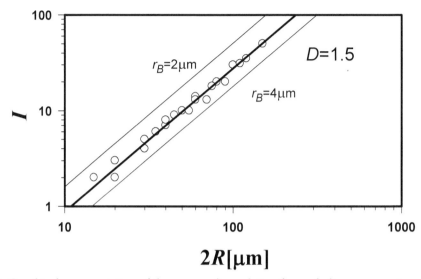

Fig. 4. Graphical representation of the mass-radius relation for asphaltene aggregates.

having the same dynamic properties, instead of the aggregate radius, makes it possible to neglect the internal permeability. For an aggregate of hydrodynamic radius r composed of $i = Ii_B$ primary particles of radius a the force balance is

$$\frac{4}{3}\pi a^3 Ii_B \left(\rho_s - \rho_f\right)g = 6\pi\eta_0 ru \tag{14}$$

Using the mass-hydrodynamic radius relations for blob and aggregate (Eqs. 9,10), one gets

$$\frac{u}{u_a} = I^{1-1/D} i_B^{1-1/D_B} \tag{15}$$

where

$$u_a = \frac{2}{9\eta_0}\left(\rho_s - \rho_f\right)ga^2 \tag{16}$$

is the Stokes falling velocity of primary particle.
Alternatively, using the expression for the hydrodynamic radius changed by the presence of blobs (Eq. 12), one obtains

$$\frac{u}{u_a} = \left(\frac{r_0}{a}\right)^{D-1} / \frac{r_0}{r} \tag{17}$$

If the blobs of the fractal dimension different from that of the aggregate are not present ($D = D_B$ and $r_0 = r$), the corresponding dependences reduce to the following relations

$$\frac{u_0}{u_a} = i^{1-1/D} \tag{18}$$

$$\frac{u_0}{u_a} = \left(\frac{r_0}{a}\right)^{D-1} \tag{19}$$

characteristic for fractal aggregates with one-level structure. Hence the following formulae

$$\frac{u}{u_0} = i_B^{1/D-1/D_B} \tag{20}$$

$$\frac{u}{u_0} = \frac{r_0}{r} \tag{21}$$

describe the free settling velocity of aggregates with mixed statistics.

5. Intrinsic viscosity of macromolecular coils and the thermal blob mass

A macromolecular coil in a solution is modeled as an aggregate with mixed statistics consisting of I thermal blobs of $D_B = 2$, each containing i_B solid monomers of radius a and mass M_a. To calculate the intrinsic viscosity

$$[\eta] \equiv \lim_{c \to 0} \frac{\eta - \eta_0}{\eta_0 c} \tag{22}$$

one has to define the mass concentration c of a macromolecular solution analyzed. The mass concentration in the coil, represented by the equivalent impermeable sphere, can be calculated as the product of the total number of non-porous monomers Ii_B multiplied by their mass $4/3\pi a^3 \rho_s$ and divided by the hydrodynamic volume of the coil $4/3\pi r^3$. This concentration multiplied by the volume fraction of equivalent aggregates φ gives the overall polymer mass concentration in the solution.

$$c = \varphi \rho_s \frac{Ii_B a^3}{r^3} \tag{23}$$

Mass-radius relations are then employed. The thermal blob mass related to that of nonporous monomer is the aggregation number of the thermal blob

$$i_B = \frac{M_B}{M_a} = \left(\frac{r_B}{a}\right)^2 \tag{24}$$

whereas the macromolecular mass related to that of thermal blob is the aggregation number of aggregate equivalent to coil

$$I = \frac{M}{M_B} = \left(\frac{r}{r_B}\right)^D \tag{25}$$

Taking into account that the volume fraction of polymer in an aggregate equivalent to polymer coil can be rearranged as follows

$$\frac{Ii_B a^3}{r^3} = I\left(\frac{r_B}{r}\right)^3 i_B\left(\frac{a}{r_B}\right)^3 \tag{26}$$

finally one gets

$$c = \varphi\rho_s\left(\frac{M_B}{M_a}\right)^{-1/2}\left(\frac{M}{M_B}\right)^{1-3/D} \tag{27}$$

or

$$c = \varphi\rho_s\left(\frac{M_B}{M_a}\right)^{-1/2}\left(\frac{M}{M_B}\right)^{-a_{MHS}} \tag{28}$$

if the fractal dimension D is replaced by the Mark-Houwink-Sakurada exponent a_{MHS}, characterizing the thermodynamic quality of the solvent, where

$$a_{MHS} = 3/D - 1 \tag{29}$$

The structure of a dissolved macromolecule depends on the interaction with solvent and other macromolecules. The resultant interaction determines whether the monomers effectively attract or repel one another. Chains in a solvent at low temperatures are in collapsed conformation due to dominance of attractive interactions between monomers (poor solvent). At high temperatures, chains swell due to dominance of repulsive interactions (good solvent). At a special intermediate temperature (the theta temperature) chains are in ideal conformations because the attractive and repulsive interactions are equal. The exponent a_{MHS} changes from $1/2$ for theta solvents to $4/5$ for good solvents, which corresponds to the fractal dimension range of from 2 to $5/3$.

The viscosity of a dispersion containing impermeable spheres present at volume fraction φ can be described by the Einstein equation (Einstein, 1956)

$$\eta = \eta_0\left(1 + \frac{5}{2}\varphi\right) \tag{30}$$

from which

$$\frac{\eta - \eta_0}{\eta_0} = \frac{5}{2}\varphi \tag{31}$$

The intrinsic viscosity can be thus calculated as

$$[\eta] \equiv \lim_{c\to 0}\frac{\eta - \eta_0}{\eta_0 c} = \lim_{c\to 0}\frac{5}{2}\frac{\varphi}{c} \tag{32}$$

Utilizing the expression for the mass concentration, one gets

$$[\eta] \equiv \lim_{c \to 0} \frac{\eta - \eta_0}{\eta_0 c} = \lim_{c \to 0} \frac{5}{2} \frac{\varphi}{c} = \lim_{c \to 0} \frac{5}{2} \frac{\varphi}{\varphi \rho_s \left(\dfrac{M_B}{M_a} \right)^{-1/2} \left(\dfrac{M}{M_B} \right)^{-a_{MHS}}} = \frac{5}{2\rho_s} \left(\frac{M_B}{M_a} \right)^{1/2} \left(\frac{M}{M_B} \right)^{a_{MHS}} \tag{33}$$

The obtained equation can be also derived in terms of complex structure aggregate parameters for any blob fractal dimension to get

$$[\eta] = \frac{5}{2\rho_s} i_B^{3/D_B - 1} I^{3/D - 1} \tag{34}$$

which is equivalent to

$$[\eta] = \frac{5}{2\rho_s} \left(\frac{r_0}{a} \right)^{3-1/D} \left(\frac{r}{r_0} \right)^3 \tag{35}$$

Equation derived for polymer coil can be compared to the empirical Mark-Houwink-Sakurada expression relating the intrinsic viscosity to the polymer molecular mass

$$[\eta] = K_\eta M^{a_{MHS}} \tag{36}$$

For the theta condition the formulae (Eq. 33) read

$$[\eta]_\theta = \frac{5}{2\rho_s} \left(\frac{M_B}{M_a} \right)^{1/2} \left(\frac{M}{M_B} \right)^{1/2} = \frac{5}{2\rho_s} \left(\frac{M}{M_a} \right)^{1/2} \tag{37}$$

and

$$[\eta]_\theta = K_\theta M^{1/2} \tag{38}$$

The Mark-Houwink-Sakurada expressions are presented in Fig. 5.

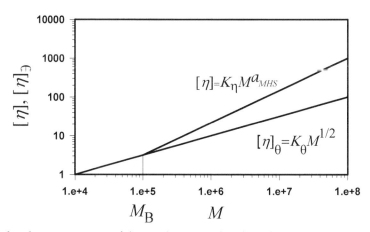

Fig. 5. Graphical representation of the Mark-Houwink-Sakurada expressions.

There is a lower limit of the Mark-Houwink-Sakurada expression applicability. Intrinsic viscosity of a given polymer in a solvent crosses over to the theta result at a molecular mass which is the thermal blob molecular mass. This means that

$$K_\eta M_B^{a_{MHS}} = K_\theta M_B^{1/2} \tag{39}$$

from which

$$M_B = \left(\frac{K_\theta}{K_\eta}\right)^{1/\left(a_{MHS}-1/2\right)} \tag{40}$$

The thermal blob mass depends on the Mark-Houwink-Sakurada constant at the theta temperature, characteristic for a given polymer-solvent system, as well as the constant and the Mark-Houwink-Sakurada exponent valid at a given temperature. The form of this dependence is strongly influenced by the mass of non-porous monomer M_a of thermal blobs, which is different for different polymers. The thermal blob mass normalized by the mass of non-porous monomer M_B / M_a, however, is the number of non-porous monomers in one thermal blob and therefore it expected to be a unique function of the solvent quality. This function, determined (Gmachowski, 2009a) from many experimental data measured for different polymer-solvent systems, reads

$$i_B = \frac{M_B}{M_a} = \left\{\exp\left[0.9 \cdot \left(2a_{MHS} - 1\right)^{1/3}\right]\right\}^{a_{MHS}/\left(a_{MHS}-0.5\right)} \tag{41}$$

The thermal blob aggregation number can be also calculated from the theoretical model of internal aggregation based o the cluster-cluster aggregation act equation (Gmachowski, 2009b)

$$i + i \sim \left[\frac{r}{R}(D)\right]^D \left(i^{1/D_i} + i^{1/D_i}\right)^D \tag{42}$$

being an extension of the mass-radius relation for single aggregate

$$i = \left(\frac{r}{a}\right)^D = \left[\frac{r}{R}(D)\right]^D \left(\frac{R}{a}\right)^D \tag{43}$$

assuming it is a result of joining to two identical sub-clusters and its radius R is proportional to the sum of hydrodynamic radii $a \cdot \left(i^{1/D_i} + i^{1/D_i}\right)$, where the normalized hydrodynamic radius is described by Eq. (5). Aggregation act equation can be specified to the form of an equality

$$i_B + i_B = 2^{1-D} \left[\frac{r}{R}(D) / \frac{r}{R}(D_{lim})\right]^D \left(i_B^{1/D_i} + i_B^{1/D_i}\right)^D \tag{44}$$

for which D tends to D_{lim} if i_B tends to infinity.

Let us imagine a coil consisting of one thermal blob. This is in fact a thermal blob of the structure of a large coil. Such rearranged blobs can join to another one to produce an object of double mass. The model makes it possible to calculate the fractal dimension D of the coil after each act of aggregation of two smaller identical coils of fractal dimension D_i changing with the aggregation progress.

Using the model for $D_{\lim} = 2$ (the fractal dimension of thermal blobs), the dependences $i_B(D)$ have been calculated using CCA simulation, starting from both good and poor solvent regions. The aggregates growing by consecutive CCA events restructured to get a limiting fractal dimension $D_{\lim}$ in an advanced stage of the process. Starting from $i_B = 8$ and $D_i = 5/3$, the result is $D=1.8115$. The second input to the model equation is thus $i_B = 16$ and $D_i = 1.8115$. Finally, the calculation results are presented in Fig. 6, where they are compared to the dependence deduced from the empirical data.

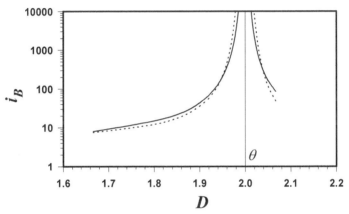

Fig. 6. Comparison of the model fractal dimension dependence of the thermal blob aggregation number (solid lines) to the representation of the experimental data measured for different polymer-solvent systems (Eq. 41), depicted as dashed lines.

6. Hydrodynamic structure of fractal aggregates

As discussed earlier, the ratio of the internal permeability and the square of aggregate radius is expected to be constant for aggregates of the same fractal dimension. Consider an early stage of aggregate growth in which the constancy of the normalized permeability is attained. At the beginning the aggregate consists of two and then several monomers. The number of pores and their size are of the order of aggregation number and monomer size, respectively. At a certain aggregation number, however, the size of new pores formed starts to be much larger than that formerly created. This means that the hydrodynamic structure building has been finished and the smaller pores become not active in the flow and can be regarded as connected to the interior of hydrodynamic blobs.

A part of the aggregate interior is effectively excluded from the fluid flow, so one can consider this part as the place of existence of impermeable objects greater than the monomers. Since both the impermeable object size and the pore size are greater than formerly, the real permeability is bigger than that calculated by a formula valid for a uniform packing of monomers. So this point can be considered as manifested by the beginning of the decrease of the normalized aggregate permeability calculated.

During the aggregate growth the number of large pores tends to a value which remains unchanged during the further aggregation. The self-similar structure exists, which can be described by an arrangement of pores and effective impermeable monomers (hydrodynamic blobs) of the size growing proportional to the pore size.

According to the above considerations one can expect effective aggregate structure such that the normalized aggregate permeability k/R^2 attains maximum. To determine the hydrodynamic structure of fractal aggregate the aggregate permeability is estimated by the Happel formula

$$\frac{k}{a^2} = \frac{2}{9\varphi} \cdot \frac{3 - 4.5\varphi^{1/3} + 4.5\varphi^{5/3} - 3\varphi^2}{3 + 2\varphi^{5/3}} \tag{45}$$

where the volume fraction of solid particles in an aggregate is described as

$$\varphi = i \cdot \left(\frac{a}{R}\right)^3 \tag{46}$$

The normalized aggregate permeability is calculated as

$$\frac{k}{R^2} = \frac{k}{a^2}\frac{a^2}{R^2} = \frac{k}{a^2}\left(\frac{i}{\varphi}\right)^{-2/3} \tag{47}$$

The results are presented in Fig. 7.

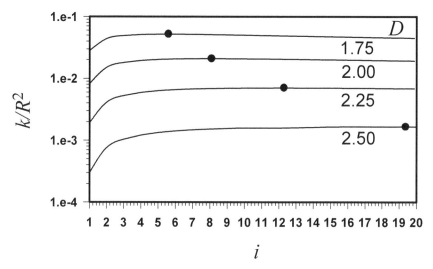

Fig. 7. Normalized aggregate permeability calculated by the Happel formula for different fractal dimensions. The maxima (indicated) determine the number of hydrodynamic blobs in aggregate.

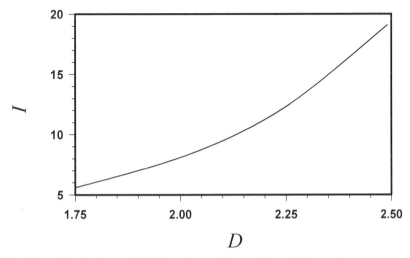

Fig. 8. Number of hydrodynamic blobs as dependent on fractal dimension.

Due to self-similarity, the number of monomers deduced from Fig. 8 is the number of hydrodynamic blobs which are the fractal aggregates similar to the whole aggregate. Hydrodynamic picture of a growing aggregate is such that after receiving a given number of monomers the number of hydrodynamic blobs becomes constant and further growth causes the increase in blob mass not their number.

As this estimation shows, the number of hydrodynamic blobs rises with the aggregate fractal dimension. The knowledge of this number makes it possible to estimate the aggregate permeability in the slip regime where the free molecular way of the molecules of the dispersing medium becomes longer than the aggregate size. In this region the dynamics of the continuum media is no longer valid.

The permeability of a homogeneous arrangement of solid particles of radius a, present at volume fraction φ, can be calculated (Brinkman, 1947) as

$$\frac{k}{a^2} = \frac{2}{9\varphi} \cdot \frac{6\pi\eta_0 a}{f_{packing}} \tag{48}$$

The friction factor of a particle in a packing can be presented as the friction factor of individual particle multiplied by a function of volume fraction of particles

$$f_{packing} = f \cdot S(\varphi) \tag{49}$$

In the continuum regime

$$f = f_{continuum} = 6\pi\eta_0 a \tag{50}$$

whereas in the slip one (Sorensen & Wang, 2000)

$$f = f_{slip} = 6\pi\eta_0 a / \left(1 + 1.612\frac{\lambda}{a}\right) \tag{51}$$

where λ is the gas mean free path.

For a given structure of arrangement (a,φ) it possible to calculate the permeability coefficient in the slip regime from that valid in the continuum regime (Gmachowski, 2010)

$$k_{slip} = k\frac{f_{continuum}}{f_{slip}} = \left(1 + 1.612\frac{\lambda}{a}\right) \cdot k \qquad (52)$$

in which the monomer size should be replaced by the hydrodynamic blob radius rising such as the growing aggregate. So large differences in permeabilities at the beginning diminish when the aggregate mass increases and disappear when the aggregate size greatly exceeds the gas mean free path.

Calculated mobility radius r_m, representing impermeable aggregate in the slip regime, is smaller than the hydrodynamic one because of higher permeability and tends to the hydrodynamic size when the difference in permeabilities becomes negligible. At an early stage of the growth of aerosol aggregates it can be approximated as a power of mass (Cai & Sorensen, 1994)

$$r_m = a \cdot i^{1/2.3} \qquad (53)$$

in which the number 2.3 greatly differs from the fractal dimension equal to 1.8.

7. Discussion

Covering the aggregate with spheres of a given size, one defines the blobs which are the units in which the monomers present in aggregates are grouped. Changing the size of the spheres we can increase or decrease the blob size. If the blobs have the same structure as the whole aggregate, the aggregate is the self-similar object.

Otherwise the object is a structure of mixed statistics with the hydrodynamic properties described in this chapter. There were analyzed aggregates containing monosized blobs of a given fractal dimension. The blobs of asphaltene aggregates are dense, probably of fractal dimension close to three. The thermal blobs - the constituents of polymer coils - have constant fractal dimension of two, independently of the thermodynamic quality of the solvent and hence the coil fractal dimension.

The determination of the hydrodynamic radius of hydrodynamic blobs in fractal aggregates, despite the same fractal structure as for the whole aggregate, serves to estimate the size of large pores through the fluid can flow. It makes it possible to model the fluid flow through the aggregate in terms of both the continuum and slip regimes.

8. References

Brinkman, H. C. (1947). A calculation of the viscosity and the sedimentation velocity for solutions of large chain molecules taking into account the hampered flow of the solvent through each chain molecule. *Proceedings of the Koninklijke Nederlandse Akademie van Wetenschappen*, Vol. 50, (1947), pp. 618-625, 821, ISSN: 0920-2250

Bushell, G. C., Yan, Y. D., Woodfield, D., Raper, J., & Amal, R. (2002). On techniques for the measurement of the mass fractal dimension of agregates. *Advances in Colloid and Interface Science*, Vol. 95, No.1, (January 2002), pp. 1-50, ISSN 0001-8686

Cai, J., & Sorensen, C. M. (1994). Diffusion of fractal aggregates in the free molecular regime. *Physical Review E*, Vol. 50, No. 5, (November 1994), pp. 3397-3400, ISSN 1539-3755

Dullien, F. A. L. (1979). *Porous media. Fluid transport and pore structure*, Academic Press, ISBN 0-12-223650-5, New York

Einstein, A. (1956). *Investigations on the Theory of the Brownian Movement*, Dover Publications, ISBN 0-486-60304-0, Mineola, New York

Gmachowski, L. (1999). Comment on „Hydrodynamic drag force exerted on a moving floc and its implication to free-settling tests" by R. M. Wu and D. J. Lee, *Wat. Res.*, 32(3), 760-768 (1998). *Water Research*, Vol. 33, No. 4, (March 1999), pp. 1114-1115, ISSN 0043- 1354

Gmachowski, L. (2000). Estimation of the dynamic size of fractal aggregates. *Colloids and Surfaces A: Physicochemical and Engineering Aspects*, Vol. 170, No. 2-3, (September 2000), pp. 209-216, ISSN 0927-7757

Gmachowski, L. (2002). Calculation of the fractal dimension of aggregates. *Colloids and Surfaces A: Physicochemical and Engineering Aspects*, Vol. 211, No. 2-3, (December 2002), pp. 197-203, ISSN 0927-7757

Gmachowski, L. (2003). Fractal aggregates and polymer coils: Dynamic properties of. In: *Encyclopedia of Surface and Colloid Science*, P. Somasundaran (Ed.), 1-10, ISBN 978-0-8493-9615-1, Marcel Dekker, New York

Gmachowski, L. (2008). Free settling of aggregates with mixed statistics. *Colloids and Surfaces A: Physicochemical and Engineering Aspects*, Vol. 315, No. 1-3, (February 2008), pp. 57-60, ISSN 0927-7757

Gmachowski, L. (2009a). Thermal blob size as determined by the intrinsic viscosity. *Polymer*, Vol. 50, No. 7, (March 2009), pp. 1621-1625, ISSN 0032-3861

Gmachowski, L. (2009b). Aggregate restructuring by internal aggregation. *Colloids and Surfaces A: Physicochemical and Engineering Aspects*, Vol. 352, No. 1-3, (December 2009), pp. 70-73, ISSN 0927-7757

Gmachowski, L. (2010). Mobility radius of fractal aggregates growing in the slip regime. *Journal of Aerosol Science*, Vol. 41, No. 12, (December 2010), pp. 1152-1158, ISSN 0021- 8502

Gmachowski, L., & Paczuski, M. (2011). Fractal dimension of asphaltene aggregates determined by turbidity. *Colloids and Surfaces A: Physicochemical and Engineering Aspects*, Vol. 384, No. 1-3, (July 2011), pp. 461-465, ISSN 0927-7757

Hausdorff, F. (1919). Dimension und äußeres Maß. *Mathematische Annalen*, Vol. 79, No. 2, (1919), pp. 157-179, ISSN 1432-1807

Sorensen, C. M., & Wang, G. M. (2000). Note on the correction for diffusion and drag in the slip regime. *Aerosol Science and Technology*, Vol. 33, No. 4, (October 2000) pp. 353-356, ISSN 0278-6826

Woodfield, D., & Bickert, G. (2001). An improved permeability model for fractal aggregates settling in creeping flow. *Water Research*, Vol. 35, No. 16, (November 2001), pp. 3801- 3806, ISSN 0043-1354

Permissions

The contributors of this book come from diverse backgrounds, making this book a truly international effort. This book will bring forth new frontiers with its revolutionizing research information and detailed analysis of the nascent developments around the world.

We would like to thank Dr. Harry Edmar Schulz, Dr. Andre Luiz Andrade Simoes and Dr. Raquel Jahara Lobosco, for lending their expertise to make the book truly unique. They have played a crucial role in the development of this book. Without their invaluable contribution this book wouldn't have been possible. They have made vital efforts to compile up to date information on the varied aspects of this subject to make this book a valuable addition to the collection of many professionals and students.

This book was conceptualized with the vision of imparting up-to-date information and advanced data in this field. To ensure the same, a matchless editorial board was set up. Every individual on the board went through rigorous rounds of assessment to prove their worth. After which they invested a large part of their time researching and compiling the most relevant data for our readers. Conferences and sessions were held from time to time between the editorial board and the contributing authors to present the data in the most comprehensible form. The editorial team has worked tirelessly to provide valuable and valid information to help people across the globe.

Every chapter published in this book has been scrutinized by our experts. Their significance has been extensively debated. The topics covered herein carry significant findings which will fuel the growth of the discipline. They may even be implemented as practical applications or may be referred to as a beginning point for another development. Chapters in this book were first published by InTech; hereby published with permission under the Creative Commons Attribution License or equivalent.

The editorial board has been involved in producing this book since its inception. They have spent rigorous hours researching and exploring the diverse topics which have resulted in the successful publishing of this book. They have passed on their knowledge of decades through this book. To expedite this challenging task, the publisher supported the team at every step. A small team of assistant editors was also appointed to further simplify the editing procedure and attain best results for the readers.

Our editorial team has been hand-picked from every corner of the world. Their multi-ethnicity adds dynamic inputs to the discussions which result in innovative outcomes. These outcomes are then further discussed with the researchers and contributors who give their valuable feedback and opinion regarding the same. The feedback is then collaborated with the researches and they are edited in a comprehensive manner to aid the understanding of the subject.

Apart from the editorial board, the designing team has also invested a significant amount of their time in understanding the subject and creating the most relevant covers. They scrutinized every image to scout for the most suitable representation of the subject and create an appropriate cover for the book.

The publishing team has been involved in this book since its early stages. They were actively engaged in every process, be it collecting the data, connecting with the contributors or procuring relevant information. The team has been an ardent support to the editorial, designing and production team. Their endless efforts to recruit the best for this project, has resulted in the accomplishment of this book. They are a veteran in the field of academics and their pool of knowledge is as vast as their experience in printing. Their expertise and guidance has proved useful at every step. Their uncompromising quality standards have made this book an exceptional effort. Their encouragement from time to time has been an inspiration for everyone.

The publisher and the editorial board hope that this book will prove to be a valuable piece of knowledge for researchers, students, practitioners and scholars across the globe.

List of Contributors

H. E. Schulz
Nucleus of Thermal Engineering and Fluids, Brazil
Department of Hydraulics and Sanitary Engineering School of Engineering of São Carlos, University of São Paulo, Brazil

G. B. Lopes Júnior, A. L. A. Simões and R. J. Lobosco
Department of Hydraulics and Sanitary Engineering School of Engineering of São Carlos, University of São Paulo, Brazil

Sergey Chivilikhin
National Research University of Information Technologies, Mechanics and Optics, Russia

Alexey Amosov
Corning Scientific Center, Corning Incorporated, Russia

German A. Maximov
N. N. Andreyev Acoustical Institute, Russia

T. L. Belyaeva
Universidad Autónoma del Estado de México, Mexico

V. N. Serkin
Benemerita Universidad Autónoma de Puebla, Mexico

V. I. Yusupov and V. M. Chudnovskii
V.I. Il`ichev Pacific Oceanological Institute, Far Eastern Branch of Russian Academy of Sciences, Russia

V. N. Bagratashvili
Institute of Laser and Information Technologies, Russian Academy of Sciences, Russia

Shinya Mizuno
Division of Virology, Department of Microbiology and Immunology, Osaka University Graduate School of Medicine, Osaka, Japan

Toshikazu Nakamura
Kringle Pharma Joint Research Division for Regenerative Drug Discovery, Center for Advanced Science and Innovation, Osaka University, Osaka, Japan

A. Montes, A. Tenorio, M. D. Gordillo, C. Pereyra and E. J. Martinez de la Ossa
Department of Chemical Engineering and Food Technology, Faculty of Science, UCA, Spain

Chiara Galletti and Elisabetta Brunazzi
Department of Chemical Engineering, Industrial Chemistry and Materials Science, University of Pisa, Italy

Laura J. Bonales, Armando Maestro, Ramón G. Rubio and Francisco Ortega
Departamento de Química Física I, Facultad de Química, Universidad Complutense, Madrid, Spain

Lech Gmachowski
Warsaw University of Technology, Poland

.

Printed in the USA
CPSIA information can be obtained
at www.ICGtesting.com
JSHW011421221024
72173JS00004B/623